DIRECTIONS IN DEVELOPMENT

W0016801

RAINS-ASIA
An Assessment Model
for Acid Deposition in Asia

Robert J. Downing
Ramesh Ramankutty
Jitendra J. Shah

The World Bank
Washington, D.C.

The findings, interpretations, and conclusions expressed in this study are entirely those of the authors and should not be attributed in any manner to the World Bank, to its affiliated organizations, or to the members of its Board of Executive Directors or the countries they represent. The boundaries, colors, denominations, and other information shown on the maps in this volume do not imply on the part of the World Bank any judgment on the legal status of any territory or the endorsement or acceptance of such boundaries.

The cover map, "Acid Deposition in Excess of Critical Loads in Asia," was generated from version 7.02 of the RAINS-ASIA model.

At the time of writing, Robert J. Downing was a consultant to the World Bank. Ramesh Ramankutty is an economist in the World Bank's Asia Technical Group. Jitendra J. Shah is an environmental engineer in the Asia Technical Group.

Library of Congress Cataloging-in-Publication Data

Downing, Robert J., 1961–
   Rains-Asia : an assessment model for acid deposition in Asia /
Robert J. Downing, Ramesh Ramankutty, Jitendra J. Shah.
      p.   cm. — (Directions in development)
   Includes bibliographical references.
   ISBN 0-8213-3919-2
   1. Acid rain—Environmental aspects—Asia—Mathematical models.
2. Energy policy—Environmental aspects—Asia—Mathematical models.
3. Energy development—Environmental aspects—Asia—Mathematical
models.   4. Asia—Economic policy—Mathematical models.
I. Ramankutty, Ramesh, 1960–   .   II. Shah, Jitendra J., 1952–   .
III. Title.   IV. Series: Directions in development (Washington,
D.C.)
QC926.57.A78D69   1997
363.738'6'011353682—dc21   97-10470
                                                                    CIP

Contents

Figures

Appendix Figure

Text Tables

Appendix Tables

Boxes

Foreword

In the past several decades, many Asian countries have experienced economic growth unmatched elsewhere in the world. Escalating demand for energy is one of the consequences of this economic growth. Although increased energy consumption indicates an improvement in the general standard of living, it also portends serious environmental consequences at the local, regional, and even global levels.

Much of the energy demand in Asia is satisfied by fossil fuels. Sulfur and nitrogen oxides are emitted by combustion of fossil fuels such as coal. These pollutants are oxidized and transported in the atmosphere. The resulting acid deposition, commonly known as "acid rain," causes severe environmental damage to natural and constructed surfaces. In addition, fine particles of sulfate and nitrate in the air can have adverse effects on human health. Acid rain knows no political or national boundaries. Its effects can be felt hundreds of kilometers from the source. Experience from Europe and North America shows that unless preventive and corrective actions are taken now, future mitigation could be quite burdensome. Waiting for the problem to become widespread before taking action will likely result in irreversible environmental damage.

As an integrated assessment tool, the RAINS-ASIA model is designed to study future energy development strategies and their implications for acid rain and to help policymakers and scientists in Asian countries explore cost-effective abatement strategies. The model allows the user to look ahead and understand what actions could be taken now to prevent future damage. RAINS-ASIA is part of a continuing effort by the World Bank and other multilateral institutions to assess the causes and effects of regional environmental problems and explore options to ameliorate them. This particular program and the associated model have been jointly funded by the World Bank, the Asian Development Bank, and several donors. Researchers and policymakers from several Asian and European countries have collaborated in its development and are currently engaged in refining and updating the model.

This report provides an overview of the model and some results of analyses that have been conducted as part of the RAINS-ASIA program. It is hoped that this report will stimulate both interest in the topic and use of the model for applications in Asia.

Mieko Nishimizu
Vice President
South Asia Regional Office
The World Bank

Jean-Michel Severino
Vice President
East Asia and the Pacific
  Regional Office
The World Bank

Acknowledgments

The RAINS-ASIA project is a collaborative effort of several research institutions in Asia, Europe, and North America. The model's development process was organized into four principal tasks: energy and emissions; transport, deposition, and monitoring; ecosystem sensitivity; and project integration. Asian and Western project leaders and focal centers were established to develop the model and to facilitate networking and information exchange among project participants. A complete list of institutions and participants appears in The Project Team, page xi.

This effort was supported with active participation from ministries and agencies in Asian countries; by grants from the Royal Norwegian Ministry of Foreign Affairs, the Norwegian Consultant Trust Funds, the Netherlands Consultant Trust Funds, the Swedish Consultant Trust Funds, and the Asian Development Bank; and with in-kind contributions from participating institutions.

The RAINS-ASIA technical report and model were peer reviewed by an international team of scientists: H. Dovland of the Norwegian Institute for Air Research, Norway; R. K. Pachauri of the Tata Energy Research Institute, India; B. Lübkert-Alcamo from the National Institute for Public Health and the Environment, the Netherlands; B. Schärer from the Federal Environmental Agency, Germany; D. M. Whelpdale from the Atmospheric Environment Service, Canada; H. Ueda from Kyushu University, Japan; K. Bull from the Institute of Terrestrial Ecology, United Kingdom; Sijin Lee from Kyong-gi University, Republic of Korea; K. C. Moon from the Korean Institute of Science and Technology, Republic of Korea; and S. Seki from the Environment Agency, Japan. Valter Angell of the Norwegian Institute of International Affairs provided valuable oversight and assistance to the project.

Special acknowledgments are due to the Internal Steering Committee at the World Bank, consisting of Anil Malhotra, Dennis Anderson, Arun Sanghvi, Todd Johnson, Carl-Heinz Mumme, Joseph Gilling, Mudassar Imran, Achilles Adamantiades, and Charles Feinstein. The task managers for the program are Jitendra J. Shah at the World Bank (with overall management guidance from Maritta Koch-Weser) and Ali Azimi at the Asian Development Bank (with overall management guidance from Bindu Lohani). Special acknowledgment is given also to the following technical contributors: Markus Amann, Gregory Carmichael, Michael Chadwick, Zhao Dianwu, Collin Green, Wesley Foell, Jean-Paul Hettelingh, and Leen Hordjik.

This Directions in Development Book is based primarily on the RAINS-ASIA technical papers prepared by the project team. The full report, "RAINS-ASIA Technical Report: The Development of an Integrated Model for Sulfur Deposition," is forthcoming from the World Bank's Asia Environmental Group. Diskettes of the RAINS-ASIA model may be ordered from the International Institution for Applied System Analysis (IIASA); the full address is provided on the last page of this book. Suhashini DeFazio, Tanvi Nagpal, and Wolf Publications were responsible for editing and producing this summary.

The Project Team

CONTRACT MANAGEMENT
World Bank
Jitendra J. Shah, Asia Technical Group

Asian Development Bank
Ali Azimi, Office of Environment

PROJECT MANAGEMENT
United States and Europe
Leen Hordijk, Wageningen Agricultural University, The Netherlands
Wesley Foell, Resource Management Associates, United States

Asia
S. C. Bhattacharya, Asian Institute of Technology, Thailand
R. M. Shrestha, Asian Institute of Technology, Thailand

ENERGY AND EMISSIONS
United States and Europe

Project leader
Wesley Foell, Resource Management Associates, United States
Institutions
Resource Management Associates, United States
Argonne National Laboratory, United States
Centre for Economic Analysis, Norway

Asia
Project leaders
S. C. Bhattacharya, Asian Institute of Technology, Thailand
R. M. Shrestha, Asian Institute of Technology, Thailand

Focal centers
Bangladesh: Bangladesh Council of Scientific and Industrial Research
China: Research Center for Eco-Environmental Sciences
India: Tata Energy Research Institute

Indonesia: Institute of Technology at Bandung
Republic of Korea: Korean Institute of Energy Economics
Japan: University of Tokyo
Malaysia: University Sains Malaysia
Myanmar: Ministry of Energy
Pakistan: Pakistan Atomic Energy Commission
Philippines: Department of Energy
Thailand: Department of Energy Development and Promotion
Vietnam: Institute of Energy

TRANSPORT, DEPOSITION, AND MONITORING
United States and Europe
Project leader
Greg Carmichael, University of Iowa, United States

Asia
Project leader
Manju Mohan, Indian Institute of Technology, New Delhi, India

Institutions
China: Research Center for Eco-Environmental Sciences
Hong Kong, China: Royal Observatory
Republic of Korea:
   Ajou University
   Korean Institute of Science and Technology
Japan:
   Central Research Institute, Electric Power Industry
   Kyushu University

Monitoring network collaborators
Bangladesh: Jahangirnagar University
China: Research Center for Eco-Environmental Sciences
China (Taiwan): Taiwan National University
Hong Kong, China: Hong Kong Polytechnic
India: Indian Institute of Technology, New Delhi
Indonesia: Institute of Technology at Bandung
Republic of Korea: Ajou University
Malaysia: Malaysia Meteorological Agency
Nepal: Nepal Meteorological Services
Sweden: Swedish Environmental Research Institute
Thailand: Environmental Research and Training Center
Vietnam: Institute of Chemistry, Center for Research

ECOSYSTEM SENSITIVITY
United States and Europe
Project leader
Jean-Paul Hettelingh, National Institute of Public Health and Environment, The Netherlands

Institutions
Stockholm Environment Institute, Sweden
University of Lund, Sweden
GEODAN, The Netherlands

Asia
Project leader
Zhao Dianwu, Research Center for Eco-Environmental Sciences, China

Institutions
Bangladesh: Jahangirnagar University
China (Taiwan): Taiwan National University
Republic of Korea: Kyong-gi University
Japan: National Institute for Environmental Studies
Vietnam: Institute of Chemistry, Center for Research

PROJECT INTEGRATION
United States and Europe
Project leader
Markus Amann, IIASA, Austria

Other participating institutions
Australia: University of Technology, Commonwealth Scientific and Industrial Research Organisation (CSIRO), Sydney
Thailand:
   King Monguts Institute of Technology
   Electric Generating Authority of Thailand
   Thailand Development Research Institute
United States:
   Oak Ridge National Laboratory
   East-West Center, Honolulu
International
   United Nations Environment Program (UNEP)
   Economic and Social Council for Asia and the Pacific (ESCAP)

Summary

Asian countries are undergoing an unprecedented economic transformation. Underlying Asia's rapid economic growth are high rates of industrialization and rapid urbanization fueled by a growing appetite for commercial energy. Demand for primary energy in Asia is expected to double every twelve years (the world average is every twenty-eight years). Fossil fuels account for about 80 percent of energy generation in Asia, with coal accounting for about 40 percent of energy produced. Because of its abundance and easy recoverability, especially in India and China, coal will remain the fuel of choice in the future. Demand for coal is projected to increase by about 6.5 percent a year, a rate that outpaces expected regional economic growth.

These trends portend a variety of environmental impacts, including acid rain caused by emissions of sulfur dioxide from burning of coal. Acid rain damages ecosystems directly and indirectly. Direct effects of acid rain include damage to foliage, particularly crop plants, whereas indirect damage occurs through acidification of soils and surface waters. At current energy consumption growth rates, by 2000 sulfur dioxide emissions from Asia will surpass the emissions of North America and Europe combined. Many ecosystems will be unable to absorb these increased acid depositions, leading to irreversible ecosystem damage with far-reaching implications for forestry, agriculture, fisheries, and tourism.

Striking similarities exist between the challenges currently facing Asia and the European situation in the late 1960s, when declining fish populations in Scandinavian countries first drew attention to the acid rain problem. Already, there is growing evidence of acid rain damage in several East Asian countries. A survey by the National Environmental Protection Agency indicates that about 40 percent of China's agricultural land is affected by acid rain. In Thailand, power production at Mae Moh using high-sulfur lignite mined in the area was responsible for serious illness among villagers living near the power plant and damage to trees and crops in the area during a 1992 episode of acid rain.

Growing concern about the acid rain problem prompted a series of expert meetings in Asia during the late 1980s. A consensus emerged that it was essential to develop an assessment tool to understand acid rain in Asia and to help develop strategies to mitigate or avert the problem. A project to develop an integrated assessment model called RAINS-ASIA (Regional Air Pollution Information and Simulation Model for Asia) emerged from this consensus. RAINS-ASIA is a computerized scientific tool to help policymakers assess and project future trends in emissions, transport, and deposition of air pollutants and their potential environmental impacts. The model was developed as an international cooperative venture involving scientists from Asia, Europe, and North America.

This book provides an overview of the RAINS-ASIA model and presents some of its results. To reach the maximum number of potential users, the model is designed to run on standard IBM-compatible computers and is user-friendly (ordering information is provided at the back of this book). A companion user's manual has been produced, and on-line help is available for guidance and troubleshooting.

Individual modules can guide users through the sequence of steps necessary for creating and evaluating emission control plans. The RAINS-ASIA model consists of three modules, each addressing a different part of the acidification process. The Regional Energy and Scenario Generator (RESGEN) module estimates energy pathways based on socioeconomic and technological assumptions; the Energy and Emission module (ENEM) uses the energy scenarios to calculate sulfur emissions and costs of control strategies; and the Deposition and Critical Loads (DEP) module calculates the levels and patterns of sulfur deposition resulting from a given scenario and then assesses the resulting environmental impacts.

In its current version, the model is designed to analyze emissions and environmental impacts of sulfur dioxide. It assesses only the indirect effects of sulfur deposition on soil. It does not include the effect of sulfur dioxide on terrestrial ecosystems through direct exposure or the effect on human health, aquatic ecosystems, and materials damage. In the future, the model and its individual modules will be validated against monitoring data.

A number of scenarios, based on assumptions about future socioeconomic conditions, have already been generated using the RAINS-ASIA model. These scenarios predict levels of energy use, emissions, and environmental pollution. The starting point of these analyses is the "base-case" or status quo scenario that forecasts future conditions assuming that no changes are made in present rates of economic and population growth or in present economic, energy, and environmental policies. In the base case, total energy demand increases at an average rate of 4 percent per year during the period 1990–2020, and the relative importance of coal in primary energy production remains comparatively stable at or near 1990 levels of 41 percent of total fuel use. Because of the high rate of economic growth forecast for the region, sulfur emissions are projected to increase from 33.6 million tons in 1990 to more than 110 million tons by 2020—an increase of 230 percent—if no actions are taken to restrict emissions.

This huge increase in energy consumption and sulfur dioxide emissions brings about similar increases in sulfur deposition. Many industrial areas of Indonesia, Malaysia, the Philippines, and Thailand experience sulfur deposition levels of 5–10 grams per square meter per year, whereas local hot spots in some industrial areas of China receive more than 18 grams of sulfur per square meter per year. In comparison, the maximum levels reached in the most heavily polluted parts of Central and Eastern Europe—the black triangle— were approximately 15 grams per square meter per year. These levels resulted in the premature death of many tree species in an area covering southwest Poland, northwest Czech Republic, and southeast Germany. The model projects that large sections of southern and eastern China, northern and eastern India, the Korean peninsula, and northern and central Thailand will receive levels of acid deposition that will exceed the carrying capacity of the ecosystem.

Although the base-case scenario may be used as the worst-case scenario (because it assumes that no new measures are undertaken to control emissions), one can also investigate the best-case scenario, of the Best Available Technology (BAT) strategy. In this scenario, sulfur dioxide emissions decrease by more than 50 percent in thirty years, from 33.6 million tons in 1990 to 16.3 million tons by 2020. As a result, nearly all areas of Asia attain sustainable levels of sulfur deposition that avoid ecosystem damage, although problems still exist in areas of China where there is heavy industrial activity. The cost of implementing the BAT strategy is estimated at US$90 billion per year, or about 0.6 percent of the region's gross domestic product (GDP).

The RAINS-ASIA model also contains an energy-efficiency scenario which assumes that concerted attempts are made to use energy more efficiently. There are a variety of control options between the extremes of the base-case scenario and the BAT scenario. The RAINS-ASIA model can simulate emissions reductions for several of these options, such as Basic Control Technology, Local Advanced Control Technology, and Advanced Emission Control Technology, and provide estimates of emission reductions and required investments for each. These reductions in emissions can cost US$2 billion–$90 billion per year (that is, up to 0.6 percent of regional GDP), based on the energy-efficiency and base-case scenarios shown in table 1. Depending on the level of ecosystem protection required for the most sensitive regions and budget limitations, the model can assist with the planning and designing of the most cost-effective options.

The RAINS-ASIA model can be used for a variety of purposes: energy and environmental planning; identifying critical ecosystems and their sulfur-carrying capacities; following emissions from an area or point

Table 1.   Emissions and Control Costs under Alternative Scenarios

Control strategies	Sulfur dioxide emissions, 2020 (millions of tons)		Control costs (billions of 1990 U.S. dollars per year)	
	Base case	Energy-efficiency case	Base case	Energy-efficiency case
Best available technologies	16	12	90	66
Advanced control technologies	50	39	39	26
Basic control technologies	63	47	40	27
No further control	111	80	4	2

source to estimate deposition; identifying the sources contributing to deposition in an ecosystem; exploring different mitigation strategies and estimating associated costs; selecting predefined energy pathways; modifying pathways to explore effects of alternative energy development strategies; and defining control strategies for individual fuel types, economic sectors, emissions control technologies, and subregions or countries.

Not only is the model a tool for analyzing air pollution effects and control strategies, it also serves an important educational function by transferring knowledge to a wide regional audience. The intended audience for the model includes planners, policymakers, and researchers concerned with energy development and environmental management issues in Asian countries, including professionals working for the government, in research organizations, in power plants, and in agricultural, soil research, and educational institutions.

This project is part of a continuing effort by the World Bank and other multilateral institutions to work with countries and regions to assess the causes and impacts of regional environmental problems and explore options for ameliorating them. It is hoped that the RAINS-ASIA model will be an important tool in this process and will help Asian countries and the World Bank evaluate the environmental consequences of development in the power and industrial sectors and adopt environmentally proactive strategies.

1
Acid Rain: An Overview

Across large parts of Asia, air pollution problems are becoming more and more evident. Rainfall in some countries, including China, Japan, and Thailand, has been measured to be ten times more acidic than unpolluted rain. Increasing evidence of acidification damage to surface waters, soils, and economically important crops is beginning to appear (see box 1). In addition, urban air quality in many areas of the region continues to deteriorate. According to United Nations Environment Programme (UNEP) estimates, twelve of the fifteen most-polluted cities in the world are in Asia, and pollution levels regularly exceed World Health Organization (WHO) guideline values by severalfold. In addition to environmental damage, these high pollution levels harm human health and have long-term regional effects on important commercial activities such as agriculture, forestry, and tourism.

Current forecasts predict continued rapid economic growth in the region. This growth will bring with it increasing emissions of air pollutants, especially sulfur. The total primary energy demand in Asia currently doubles every twelve years (compared with a world average of every twenty-eight years). Approximately 80 percent of energy in Asia is produced by burning fossil fuels, and biomass supplies an additional 15 percent. Coal is expected to continue to be the dominant energy source, with demand projected to increase by 6.5 percent per year, a rate that outpaces regional economic growth. If current trends in economic development and energy use in Asia continue, emissions of sulfur dioxide, one of the critical components in acid rain, will more than triple within the next thirty years. Many ecosystems will be unable to continue to absorb these increased levels of pollution without harmful effects, thus creating a potential danger for irreversible environmental damage in many areas.

Box 1. Acid Rain Damage in Asia

Bangladesh

- Monitoring data from two sites in the Dhaka area show increased acidity in winter rainfall (Ahmad 1991) and large increases in sulfate concentrations.

China

- A survey of forest growth in the Sichuan Basin and Guizhou province has indicated that the incidence of forest damage (indicated by rates of tree and foliage growth) is higher in areas with highly acidic rain (Dianwu and Xiaoshan 1992). Areas receiving rainfall with a pH of 4.5 or less also show higher rates of material corrosion in exposure tests.
- A study of the southwestern provinces of Sichuan and Guizhou has calculated that approximately two-thirds of the agricultural land in the area received some amount of acid rain (pH value of less than 5.0), with a total of 16 percent of the crop area damaged to some extent by acid rain. A drop in crop yields caused by acid rain is expected to cost between 17 million and 27 million yuan, or about 0.5 to 0.8 percent of the total agricultural output value from the region (Dianwu and Xiaoshan 1992).

India

- Measurements of rain chemistry show increasing acidity at most monitoring sites in India (Sridharan and Saksena 1991). Because the soils in the northeastern part of the country are already acidic, they are more prone to ecological damage from increasingly acidic rain.

Republic of Korea

- Ten years of monitoring data from the Seoul National University show that rainfall is nearly always more acidic than unpolluted rain and often is at least ten times more acidic (pH of 4.2 to 4.4). The most highly polluted rain or snow occurs during winter, when home heating contributes to peak sulfur emissions (Hong 1991). Laboratory experiments have shown damaging effects on important plant and tree species at these levels of acidity.
- A comparative study of growth rates of pine and oak trees (reported in Hong 1991) in urban and rural areas showed that soil acidity levels in urban areas are up to ten times those of similar rural sites. These urban sites also have higher concentrations of toxic metals such as magnesium. Growth rates in both areas have declined significantly since 1970.

Vietnam

- Monitoring data (reported by Gian, Van, and Nihn 1992) show sulfur dioxide concentrations of up to 500 micrograms per cubic meter in some industrial and urban areas. In Ho Chi Minh City, fuel combustion is responsible for 98 percent of the 42,000 tons of sulfur dioxide emitted each year.

The Acidification Phenomenon

Acid rain is the product of chemical reactions between airborne pollution (sulfur and nitrogen compounds) and atmospheric water and oxygen. Once in the atmosphere, sulfur dioxide (SO_2) and nitrogen oxides (NO_x) react with other chemicals to form sulfuric and nitric acids. These substances can stay in the atmosphere for several days and travel hundreds or thousands of kilometers before falling back to the earth's surface as acid rain. This process is more accurately termed "acid deposition," because acidity can travel to the earth's surface in many forms: rain, snow, fog, dew, particles (dry deposition), or aerosol gases. A simplified overview of the process is shown in figure 1.

Although sulfur and nitrogen compounds can be generated by biological processes such as natural soil decomposition or other natural sources such as volcanoes, most sulfur emitted into the atmosphere results from anthropogenic (of human origin) activities. Coal- and oil-fired power-generating stations, domestic heating, biomass burning, various industrial processes, and transportation are all important sources of emissions that cause acid deposition.

Figure 1.   Processes Involved in Acid Deposition

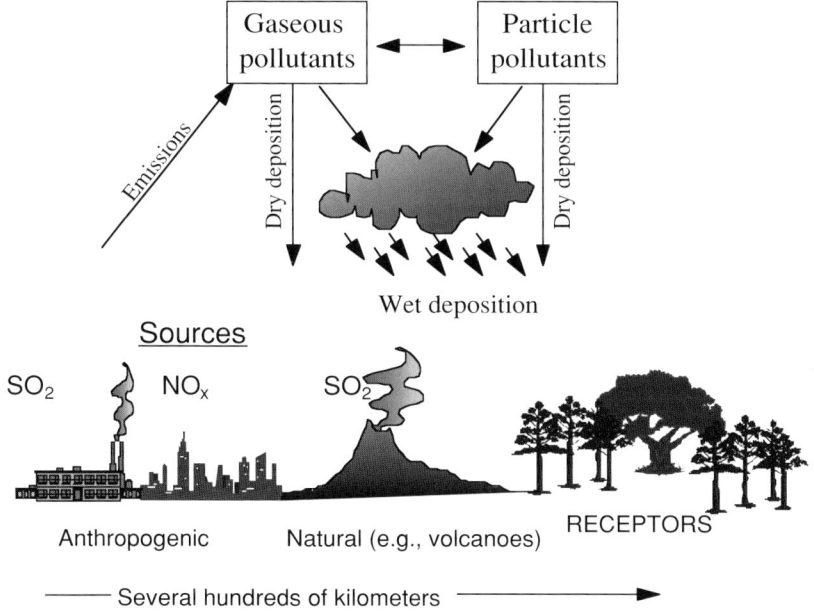

The documented effects of air pollution and acid deposition include the following:

- Major contributions to forest decline, possibly in complex interactions with natural stresses
- Release of toxic metals such as aluminum that can damage soils, vegetation, and surface waters
- Direct damage to crops and vegetation by high air concentrations of pollutants or indirect damage through chemical changes in the soil
- Damage to aquatic resources and their ecosystems
- Increased rate of erosion of monuments, buildings, and other cultural and commercial resources
- Direct, adverse effects on human health, especially for sensitive populations with respiratory or cardiovascular problems.

Emissions from large point sources of sulfur emissions such as power plants were once considered a local problem. As awareness of the harmful effects of these pollutants grew, however, new facilities were built with taller smokestacks, designed to spread the pollution over a larger area. This wide dispersion makes the long-range acidification problem and its possible solutions a national and regional concern.

The Integrated Assessment Approach to Acid Deposition

To adequately understand an issue as complex as the acidification process on a continental scale, it is necessary to develop a comprehensive framework that analyzes all the major components of the process to adequately identify links among various sources, processes, and effects and to design effective strategies for ameliorating the problem. This framework must address a variety of often conflicting considerations such as the following:

- Future trends in economic growth in various regions
- Policy options to meet the resulting increases in energy demand
- Emissions of air pollutants, available control options, and costs of reducing them
- Geographic dispersion of emissions by atmospheric transport and deposition
- Locations and response to increased pollution levels of a variety of sensitive ecosystems (including agricultural crops, forests, surface waters, materials, and human health)
- Economic and environmental consequences of implementing emissions control strategies.

Integrated assessment models such as RAINS-ASIA provide insight into these issues by establishing interrelationships and links among various aspects of the acidification process. The model provides analysis and comparison of the costs and benefits of various options to reduce the regional effects of air pollution before they occur and cause irreversible damage.

Learning from the European and North American Experience

There are striking parallels between the challenges that Asian countries currently face and the development of coordinated international responses to similar environmental threats in Europe and North America. In the late 1960s, marked declines in fish populations were noted in many Scandinavian countries. Scientists concluded that precipitation over these countries was gradually becoming more acidic as a result of sulfur emissions in other parts of Europe. Acid rain was also suspected of being a primary cause of soil acidification and the resulting forest dieback in some parts of Central Europe. By the mid-1980s, Germany observed that more than half of its forests showed some effects from air pollution, furthering the impetus toward a coordinated international response to the worsening issue. Increasing concerns about air pollution being transported over long distances led to the signing of an international agreement in 1979. This agreement committed European and North American countries to attempt to limit or reduce their transboundary emissions of air pollutants (see box 2).

In North America, the effects of acid deposition on the environment and human health became an issue of increasing public concern during the 1970s. Canada's investigation into the possible causes of acidified lakes and damaged forest areas led to the conclusion that increasing levels of sulfur emissions, primarily from the northeastern United States, combined with the use of tall smokestacks (designed to disperse pollution from the immediate surroundings of large pollution sources), were contributing to this ecosystem damage hundreds—perhaps thousands—of kilometers away. Since the early 1980s, Canada has defined quantitative environmental objectives, in terms of target levels of sulfur deposition, aimed at protecting sensitive ecosystems from acidification. In 1985, Canada established a program to reduce sulfur emissions in ecologically sensitive regions in eastern Canada to a maximum of 2.3 million metric tons by 1994.

In the 1980s, in response to concerns about possible acidification damage to lakes, streams, and forests in the sensitive northeast region, the United States began large-scale research and monitoring programs to study

Box 2.   The Convention on Long-Range Transboundary Air Pollution and the RAINS-ASIA model

The 1979 Convention on Long-Range Transboundary Air Pollution, the first legally binding international agreement dealing with air pollution on a regional basis, also provided a mechanism for international cooperation on research and monitoring activities. The scope of the convention, established under the auspices of the United Nations Economic Commission for Europe (UN ECE), is broad: its 38 members include most countries of Eastern and Western Europe, the United States, and Canada. The convention provides a unique forum for wide-ranging cooperation on a broad range of air pollution issues.

To provide a scientific basis for negotiating additional agreements to limit specific pollutants, the convention established international cooperative monitoring, research, and assessment activities. As a result of these efforts, protocols to reduce sulfur dioxide (1985), nitrogen oxides (1988), and volatile organic compounds (1991) have already been agreed on in the ECE region. These agreements, however, are based solely on individual countries' voluntary commitments to reduce emissions by a specific percentage target. No concerted attempt was made to try to maximize the environmental benefits of these emissions reductions.

Beginning in the early 1980s, an international team of scientists came together at the International Institute for Applied Systems Analysis (IIASA) in Austria to begin development of a computer model to analyze the sources and effects of acid deposition in Europe, and possible strategies for its control. This multidisciplinary group developed the RAINS model—an integrated assessment tool that addresses all aspects of the acidification phenomenon, including energy use, emissions, control costs, pollutant transport and deposition, and environmental effects. Widespread international acceptance of the RAINS model led to its use as a principal scientific support tool in developing a new protocol to further reduce sulfur emissions in the UN ECE region. The new international agreement, signed in Oslo in 1994 by thirty-three countries, is unique in its effects-based strategy. It takes into account the ability of the environment to withstand pollution, while assigning each country a different emissions-reduction target. Together with prescribing national emission ceilings, the protocol contains provisions on the joint implementation of emission reductions to make abatement measures more cost-effective. In addition, the protocol contains requirements for the use of Best Available Technology (BAT) in new large combustion plants, gas and oil sulfur content limits, and provisions for exchange of technology.

the causes and effects of acid deposition and options for controlling it. The National Acid Precipitation Assessment Program and a five-year, $5 billion Clean Coal Technology Program documented the consequences of and possible solutions to the acid deposition problem. In 1990, the U.S. Congress passed major amendments to its Clean Air Act, mandating a 10-million-ton reduction in sulfur emissions by 2000. In March 1991, the

United States and Canada signed a bilateral agreement on transboundary air quality issues that establishes specific targets and timetables for reducing sulfur dioxide and nitrogen oxides and improves coordination on a variety of joint research and monitoring programs.

The Outlook for Asia

Europe and North America have provided a clear lesson: Continued growth of fossil fuel use without any abatement measures will lead to significant environmental damage. Figure 2 compares projected trends in sulfur dioxide emissions in Asia, Europe, and North America. The expected results of international action to reduce sulfur emissions in both Europe and North America are readily apparent because sulfur emissions in both regions are expected to decrease significantly in coming years. In contrast, if present energy and environmental policies remain unchanged, rapid economic development in Asia will lead to an unprecedented increase in sulfur emissions in the region, to a total of more than 110 million tons of sulfur dioxide by 2020.

Figure 2.   Past and Projected Sulfur Dioxide Emissions for Asia, Europe, and the United States and Canada

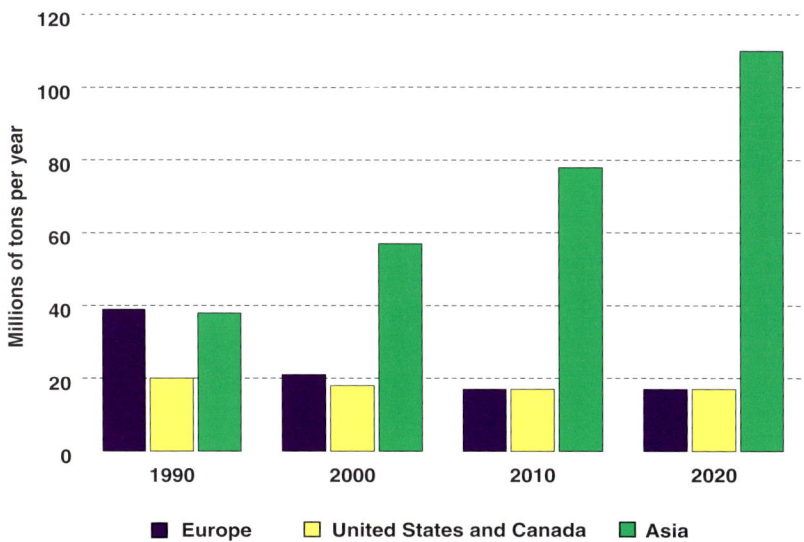

Note: It is assumed that emissions are stabilized in Europe and North America by 2010.
Sources: For Europe, Cofala and Schopp (1995). For United States and Canada, National Acid Precipitation Assessment Program (1991). For Asia, Reference Scenario from RAINS-ASIA model.

In light of these disturbing trends, it is vitally important to apply the lessons learned in the West to address the air pollution problems in Asia. Experience in addressing large-scale environmental problems has shown that it is much cheaper to implement measures to *prevent* environmental hazards than to clean up the pollution once it has occurred.

This proactive method of environmental management, using tools such as the RAINS-ASIA model, helps to ensure that the most effective and efficient policies can be identified and implemented. As was the case in Europe and North America, the broad geographic nature of the problem requires *national and regional* problem solving as a key to developing workable, long-term strategies to reduce or prevent the environmental effects of air pollution.

Current Acid Rain–Related Monitoring Activities in Asia

Asian countries are now beginning to establish monitoring programs and regulatory measures for acid rain. A network that measures sulfur dioxide concentrations emerged in Phase I of RAINS-ASIA. This network and another that monitors acid deposition in Asia are described below (see also table 2).

Table 2. Summary of Sulfur Dioxide–Monitoring Data from the Passive Sampling Network
(monthly averages, January 1994–February 1995, for stations reporting)

Country	Number of stations reporting data	Annual mean SO_2 concentration (micrograms per cubic meter)
Bangladesh	2	6.4
China	5	22.5
Hong Kong, China	1	24.0
India	6	6.5
Indonesia	4	1.2
Korea, Rep. of	3	5.0
Malaysia	4	0.9
Nepal	7	2.7
Taiwan, China	1	1.3
Thailand	5	3.5
Vietnam	2	14.2
Total	40	

Note: Not all stations report data.
Source: Carmichael and Arndt (1995).

Figure 3. Locations of Passive Sulfur Dioxide Samplers Established in the RAINS-ASIA Project

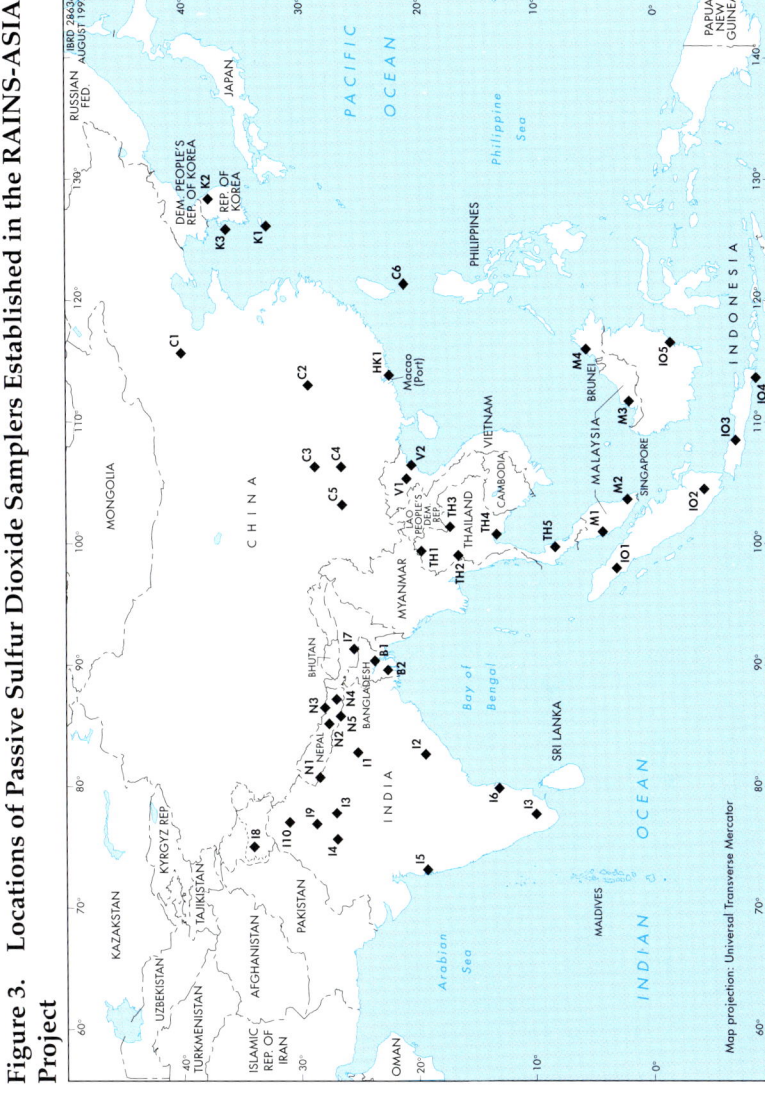

Source: Carmichael and Arndt (1995).

Sulfur Dioxide–Monitoring Network

High levels of sulfur dioxide, a principal contributor to acid rain, are being measured at an increasing number of locations. A notable result from Phase I of this project has been the initiation of a network of inexpensive sulfur dioxide air samplers at forty-three sites in eleven countries to obtain more broadly based monitoring data for the model. Because most existing sulfur dioxide–monitoring stations in the region are based in urban areas, it was necessary to gather more base-level data from rural sites for further model evaluation.

The monitoring network is designed to provide information on regional, long-term, monthly, and annual average sulfur dioxide concentrations. The sites selected are situated away from large emission sources and in regions expected to be highly sensitive to acid deposition. The first results from the network are presented in table 2, and the locations of the sites are shown in figure 3. (Note that not all stations are yet operational and reporting data.)

Acid Deposition–Monitoring Network in East Asia

Administrators and scientists from several East Asian countries (China, Indonesia, Japan, Korea, Mongolia, the Philippines, and Thailand), as well as the Russian Federation, are developing a cooperative initiative under the leadership of the Environment Agency of Japan to establish an Acid Deposition–Monitoring Network in East Asia. Guidelines have been developed for monitoring acid deposition in the East Asia region, with five principal components: (a) establishment of national acid deposition monitoring networks; (b) establishment of a center (or centers) for the monitoring network in the region; (c) exchange of data, experience, and information among participating countries; (d) central compilation and analysis of monitoring data; and (e) capacity-building activities. The network is also expected to collaborate with programs such as RAINS-ASIA. Country representatives have met on three occasions to discuss the arrangements for establishing the network. The Fourth Expert Meeting was held in Hiroshima, Japan, from February 4 to February 6, 1997. At this meeting, the participants established technical manuals for monitoring acid deposition and developed an overall schedule for establishing an acid monitoring network in East Asia.

2
Institutional Arrangements for the RAINS-ASIA Program

Growing awareness of the magnitude of current and projected air pollution problems in Asia prompted the organization of the first international symposium on "Acid Rain and Emissions in Asia" in 1989. The meeting was organized by the Asian Institute of Technology (AIT), Argonne National Laboratories, and Resource Management Associates and was held at AIT in Bangkok, Thailand. The symposium was the first in a series designed to bring together experts from Asia, Europe, and North America to assess present and possible future energy use, sulfur emissions, and environmental risks from long-range transboundary air pollution (Foell and Green 1992; Foell and Sharma 1991).

Based on existing evidence of environmental damage in some heavily industrialized areas and the ramifications of the rapid growth predicted in fossil-fuel use throughout Asia during the next thirty years, the meeting recommended the following actions:

- Establish an intensified monitoring program to assess the current pollution problem in Asia.
- Develop a coordinated research effort on the atmospheric transport and deposition of pollutants in Asia and their effects on natural and constructed systems.
- Initiate an integrative program of assessment and policy analysis to analyze long-term strategies for acid deposition problems on national and regional scales.

Participants decided to build on the successful implementation of RAINS in Europe and use it as a tool to analyze long-term trends, strategies, and options for air pollution problems on different geographic scales. The model's integrated method and framework have given it widespread acceptance as a decisionmaking tool.

An important aspect in the development of the RAINS model is cooperation among various international, regional, and national institutions. The program was supported by various grants and in-kind contributions. From the beginning of the project, the importance of involving regional experts in the structure, design, and implementation of the model was stressed. To this end, three networks of Asian scientific and research institutions were established to facilitate the collection and review of input data, assist in defining realistic future scenarios for the region, and provide advice on the modifications necessary to adapt the RAINS framework to the Asian situation. Close collaboration has been maintained throughout the project with participating institutes in Europe and the United States through workshops and meetings, and there has been continuous communication during the model development process.

Networks

The following networks were established during this project:

- *Energy and emissions.* A major task in Phase I of the project was the establishment of a network of Asian energy research institutions to collaborate in developing regional databases and energy scenarios. With funding from the Asian Development Bank, the Asian Institute of Technology (AIT) in Bangkok was designated as a coordinating center for the network (with national contact persons from twelve countries). AIT was responsible for collecting the necessary data on a national and regional level and establishing long-term ties to principal institutions and potential model users.
- *Atmospheric transport and deposition.* The Center for Atmospheric Sciences, Indian Institute of Technology (IIT) in New Delhi, served as the principal hub of communication and activities related to the development of the atmospheric transport and deposition part of the model. Phase I activities related to the center included installation of computer hardware and software to run the RAINS-ASIA model at the center; training of IIT research staff and scientists from China, India, Indonesia, Japan, and Korea in the use of the models; and development of an international network of scientific researchers and other contact persons to provide progress reports and technical consultations concerning the operation of sulfur dioxide monitoring sites.
- *Environmental impacts.* The Research Center for Eco-Environmental Sciences in Beijing served as a focal point for the collection of national input data on ecosystem effects. Together with a network of environmental researchers from several Asian nations and Western collaborating institutions, researchers participated in numerous workshops during Phase I of the project to plan and review research methods and results. A geographic information system (GIS) and

critical load models were installed at the center, and training in the systems was provided through periodic exchanges of personnel. These activities could be expanded to include other nations and institutions in a future phase of the project.

Institutions participating in each of these networks are listed in the Acknowledgments and The Project Team sections at the beginning of this book. Significant efforts have also been devoted to integrating the work of the various groups, including developing links among the computer modules, in the design of the model interface, and production of a user's manual. IIASA in Vienna has coordinated training sessions and annual conferences to review the progress of the project. In addition, the World Bank has sponsored various coordination and review meetings.

3
The RAINS-ASIA Model

The RAINS-ASIA model consists of various modules, each of which addresses a different part of the air pollution and acidification process. A simplified overview of the links among the model's components (or modules) is shown in figure 4. Each of these modules is more fully described in this chapter:

- The Regional Energy and Scenario Generator (RESGEN) module estimates energy consumption patterns based on socioeconomic and technological assumptions.
- The Energy and Emissions (ENEM) module uses these energy scenarios to calculate sulfur emissions and the costs of selected control strategies.
- The Deposition and Critical Loads (DEP) module, which consists of the Atmospheric Transport and Deposition (ATMOS) submodule and the Environmental Impact and Critical Loads (IMPACT) submodule, calculates the levels and patterns of sulfur deposition resulting from a given emissions scenario and the ecosystem critical loads and their environmental impacts based on these patterns.

To create a practical tool for scenario analysis, the RAINS-ASIA model defines simplified relationships between input data (for example, economic development in the RESGEN module, annual emissions in the ATMOS module, and deposition in the IMPACT module) and the output variables (for example, annual emissions in the ENEM module, deposition in the ATMOS module, and potential ecosystem damage in the IMPACT module). The model uses these relationships to develop an overall assessment framework, allowing for the comparative analysis of alternative energy and emissions reduction strategies.

The RAINS model can answer many policy-relevant questions, including the following:

- How do various economic and development policies affect energy production and consumption patterns and the resulting emissions and deposition levels?

Figure 4.   Major Components of the RAINS-ASIA Model

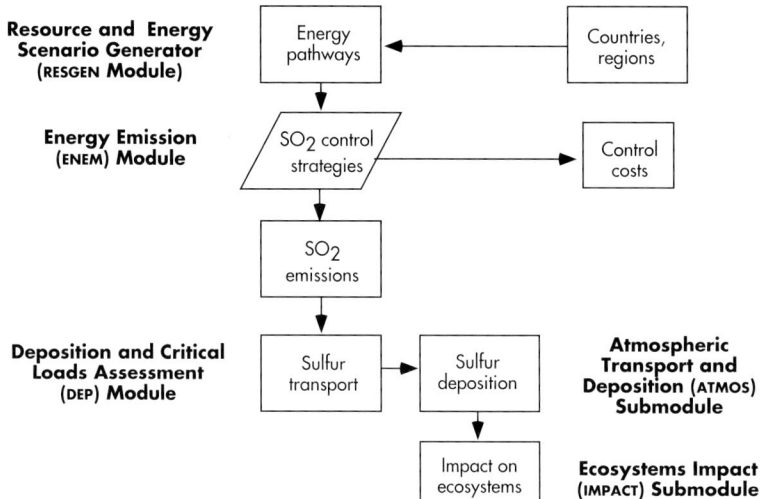

- How do energy policies such as increasing energy efficiency, switching to fuels with lower sulfur content, or implementing emissions control measures affect energy demand and emissions?
- How do changes in the spatial distribution of emissions sources change emission and deposition patterns?

Scope and Limitations of the Model

The region covered by the model ranges from 10° south to 55° north latitude and from 60° to 150° east longitude, covering the countries of East, South, and Southeast Asia. The model uses 1990 data as a base and calculates future energy, emissions, and environmental parameters through 2020, in ten-year increments.

Although the model's geographic scope is broad, it is also detailed, covering a total of ninety-four separate regions in twenty-one countries (see table A1). Twenty-two of these regions are major metropolitan areas, and international sea lanes constitute one region. The model's databases also include information (see appendix for details) on 6 end-use energy consumption sectors, 17 fuel types, 355 large point sources of sulfur dioxide emissions, and 31 ecosystem types.

Although the initial version of the model provides a general view of the acidification problem in Asia, certain limitations are unavoidable. Most notably, the initial implementation focuses mainly on the potential sulfur acidification problem in Asia and excludes other important air-pollution-related problems such as urban air quality, global climate change, and tropospheric ozone. Future refinements of the model will address the contributions made by nitrogen oxides and ammonia as well as other pollutant species (carbon dioxide, ozone, particulate matter, and volatile organic compounds).

The model addresses many facets of the acidification problem, but it restricts itself to the description of the major physical flows of air pollutants in the biosphere. Only the soil-based effects of acid rain have been incorporated for now. Thus, the effects on rivers, lakes, materials, and human health will have to be incorporated in a future version of the model. In addition, many economic aspects of regional air pollution problems, such as the potential role of economic instruments for reducing emissions or the economic value of avoiding environmental damage, have not been incorporated into the model. Complex links between regional environmental issues and global concerns such as climate change have also not been addressed as yet.

The Regional Energy Scenario Generator (RESGEN) Module

The RESGEN module estimates present and future energy supply and consumption levels based on a variety of socioeconomic and technological assumptions. Given a set of specifications concerning current and future conditions (using either the extensive socioeconomic and energy demand and supply databases in the model or user-specified alternative assumptions), the model calculates energy scenarios for the period 1990–2020. These energy scenarios can then be used as an input into the RAINS model to calculate sulfur dioxide emissions, deposition, and environmental effects.

The RESGEN module is structured to provide answers to policy-relevant questions such as the following:

- What are the effects of changes in population and economic growth on future energy demand?
- How do economic and development policies for various economic sectors affect energy production and consumption patterns?

Structure

The following steps are required to calculate energy consumption levels and patterns for a given scenario:

- Assumptions about future socioeconomic information—population and GDP growth rates—for each country are used to calculate total national energy demand, allocated among six economic sectors.
- These national energy demand figures are apportioned among national subregions (if any), based on current socioeconomic data and projected trends.
- Energy demand for each sector and region is disaggregated into seventeen fuel types based on current fuel use information and projected trends.
- National energy supply requirements are calculated based on the results of the energy demand calculations.
- Existing and planned power plants, including information on plant size and type(s) of fuel used, are identified.
- Remaining national energy demand is apportioned to smaller sources whose regional distribution and fuel(s) used are defined by the model's user.

These steps result in a scenario that describes total energy demand by subnational region, economic sector, and fuel type.

With RESGEN, a user can select, review, and modify the following critical parameters on the subcountry (regional) level:

- *Socioeconomic data,* including rates of population growth and growth of gross domestic product (broken down into three components: industrial, agricultural, and commercial/other).
- *Growth rates of energy demand* among industrial, transportation, residential, commercial, and agricultural sectors and other uses. Energy supply and transformation systems are subdivided into electricity generation, oil refining, and other industrial operations.
- *Energy demand* per unit of economic activity for each of the six end-use sectors.
- *Fuel types used.* Sulfur dioxide emissions are highly dependent on the type and characteristics of the fuel used; therefore, the model considers seventeen fuel types, including various qualities of coal, other solid fuels, fuel oil, natural gas, renewable sources, hydropower, and nuclear power (see table A3).
- *Fuel characteristics* such as the sulfur content of various fuels.

Two energy-demand scenarios have been developed to describe possible future energy pathways for Asia. These scenarios provide a yardstick by which to measure the effects of various energy policies and control strategies. The base-case scenario relies on official energy projections from individual countries whenever available. Business-as-usual policies that use historical or expected trends were assumed in areas for which future economic or energy data are incomplete or unavailable.

The energy-efficiency scenario, in contrast, assumes the introduction of policies to improve energy efficiency and to shift from using high-polluting fuels to using low-sulfur fuels ("fuel substitution"), especially in the power-generation sector. This scenario demonstrates the importance and potential of energy-efficiency and fuel substitution measures to reduce emissions.

Principal Results

Total energy demand, by fuel type, for the base-case and "energy-efficiency" scenarios is shown in figure 5. In the base-case scenario, energy consumption would increase at an average rate of 4 percent during the thirty-year period from 1990 to 2020 compared with a growth rate of 3.1 percent under the energy-efficiency scenario. Total consumption levels more than triple in the base-case scenario, from a 1990 level of 83.5 exajoules to more than 274 exajoules by 2020.

Although the energy-efficiency scenario reflects the widespread application of measures to increase energy efficiency, energy growth in the region is still forecast to more than double during the same thirty-year period.

In the base-case scenario, the relative importance of coal in primary energy production would remain stable at or near 1990 levels of 41 percent of total fuel use. The use of natural gas would increase fivefold, reaching a level of 9 percent of total primary energy by the end of the period. Although total levels of biomass fuels remain relatively constant, their share of energy consumption would decrease from 15 percent in 1990 to 8 percent in 2020.

Conversely, the energy-efficiency scenario shows a 31 percent reduction in coal usage from base-case levels by 2020. This reduction is a result of the combined effects of improved energy efficiency and fuel substitution in the power-generating sector, reflecting a move away from coal to lower emission fuels such as hydropower, nuclear energy, and natural gas.

Energy and Emissions (ENEM) Module

The ENEM module of the RAINS-ASIA model uses the information on energy demand, types of fuels used, and location of major emission sources developed by the RESGEN module and estimates the resulting amounts

---

A "joule" is a standard international unit of energy that is equivalent to 1 watt for 1 second. An exajoule (EJ) is equal to 10^{18} joules, or 1 billion billion joules.

Figure 5.   Total Energy Demand, by Fuel Type, for the Base-Case and Energy-Efficiency Scenarios

Base-Case Scenario

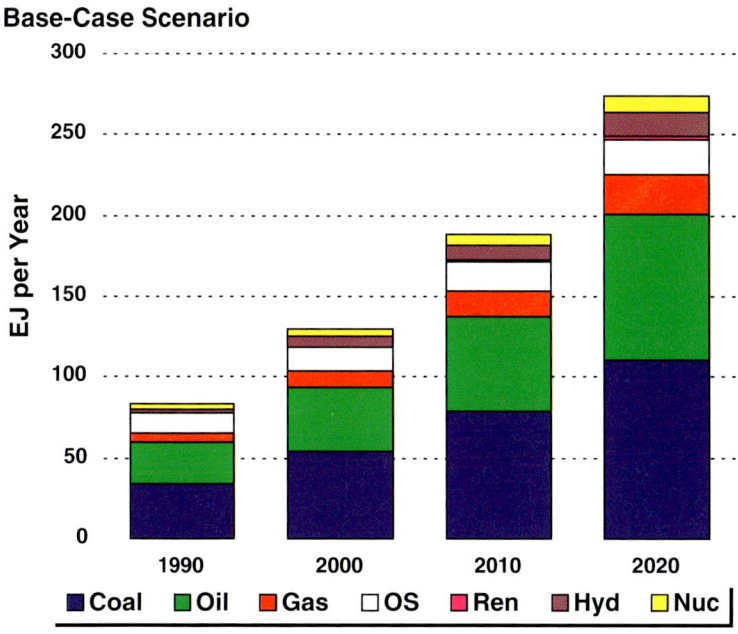

Energy Efficiency Scenario

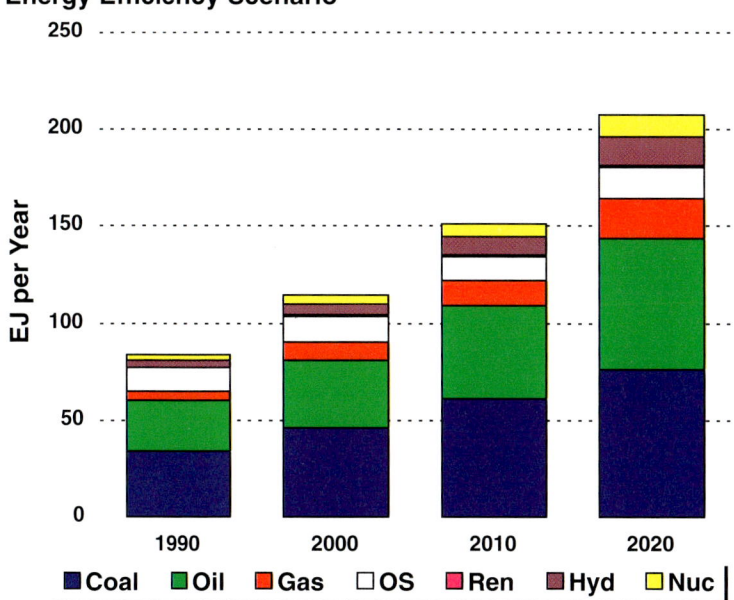

and patterns of sulfur dioxide emissions and the costs of various control options.

The ENEM module contains energy and emissions databases (developed primarily from in-country data sources) covering 94 regions in 21 countries, including 355 individual large point sources such as existing or planned power plants or industrial sites. It allows users to choose future energy scenarios, calculate the resulting emissions levels, and answer various policy-related questions such as these:

- What will future levels of sulfur emissions be for a given economic or energy scenario?
- What control strategies can be used to reduce sulfur emissions, and what are the estimated costs of implementing these controls?
- What effect do changes in energy efficiencies have on sulfur emissions?
- What environmental improvements would result from the relocation of a large emission source from a sensitive ecological area to a less sensitive region?

Although the necessary data were available for some countries from previously published studies, for many countries this information had to be developed from secondary sources. Data on energy use by fuel type and geographic region, the size and location of large point sources, and fuel characteristics (sulfur content and emission factors) for various fuels were compiled by the project team. The model includes emissions data developed specifically for this project from seventeen countries and previously published emissions estimates for six other countries. The network of energy research institutions was instrumental in gathering these data. With the use of this network, the first Asia-wide databases on fuel consumption by region and economic sector, fuel characteristics, and emissions control options and costs were created and incorporated into the RAINS-ASIA model.

Structure

The ENEM module required the creation of a grid-based base-year emissions inventory and a review of control technologies (see table A4). The base-year inventory includes all anthropogenic (artificial) sources of sulfur emissions, both from land-based sources and from major international shipping lanes. Also included are emissions from biomass burning (agricultural and animal wastes and wood used as fuel) and from volcanic eruptions.

The base-year emissions inventory developed in Phase I served as a basis for estimating future emissions levels under various energy scenarios. This process involved data collection, for all regions and large point sources considered in the model, on sulfur content and heating values of fuels and on the fraction of sulfur retained in ash after combustion (for solid fuels). Regional or site-specific data on fuel characteristics were used wherever available. In a few cases in which no national data were available, values of similar parameters from the European version of the RAINS model were used (see box 3). These values may be substituted by more suitable Asia-specific data when they become available. A continuing process of review and feedback on data and sources, involving all member institutions in the energy network, was used to improve the quality of the final databases. The locations of 355 large point sources used in the model are shown in figure 6.

Box 3.   Emissions Control Technologies and Costs

With ENEM, the model's user can investigate a number of emissions control options focusing on reducing the sulfur contained in fuel before, during, or after combustion. The user can select emissions control techniques to be applied to particular large point sources, in specific economic sectors, or in certain geographic regions. The following control measures are considered:

- Use of low-sulfur hard coal, either from naturally occurring low-sulfur coal types or by some degree of coal washing
- Use of low-sulfur heavy fuel oil, either from low-sulfur crude or oil desulfurized during refining
- Use of diesel oil (gas oil) with lower sulfur content
- Desulfurization during the combustion process (for example, through limestone injection or fluidized bed combustion processes)
- Desulfurization of flue gas after combustion.

The ENEM module also performs cost calculations for implementing the selected emissions-reduction strategies. The model calculates total life-cycle costs, including investment-related start-up costs (installation, construction, and working capital), fixed operating costs (maintenance, taxes, and overhead), and variable operating and maintenance costs (additional labor and waste disposal). The parameters used in the calculations are determined to be either common or specific to a particular country. Common parameters, which apply to all instances of a specific technology, include installation lifetimes, sulfur removal efficiency, and energy and material requirements. Country-specific parameters include items such as interest rates, average plant capacity utilization, boiler and furnace size, and energy and material prices.

The module also produces national cost curves that rank the available abatement measures in terms of their overall cost-effectiveness. Because factors such as energy use patterns and technological infrastructure differ greatly among countries in the region, there are large differences in national cost curves.

Figure 6. Locations of the 355 Large Point Sources in the RAINS-ASIA Model

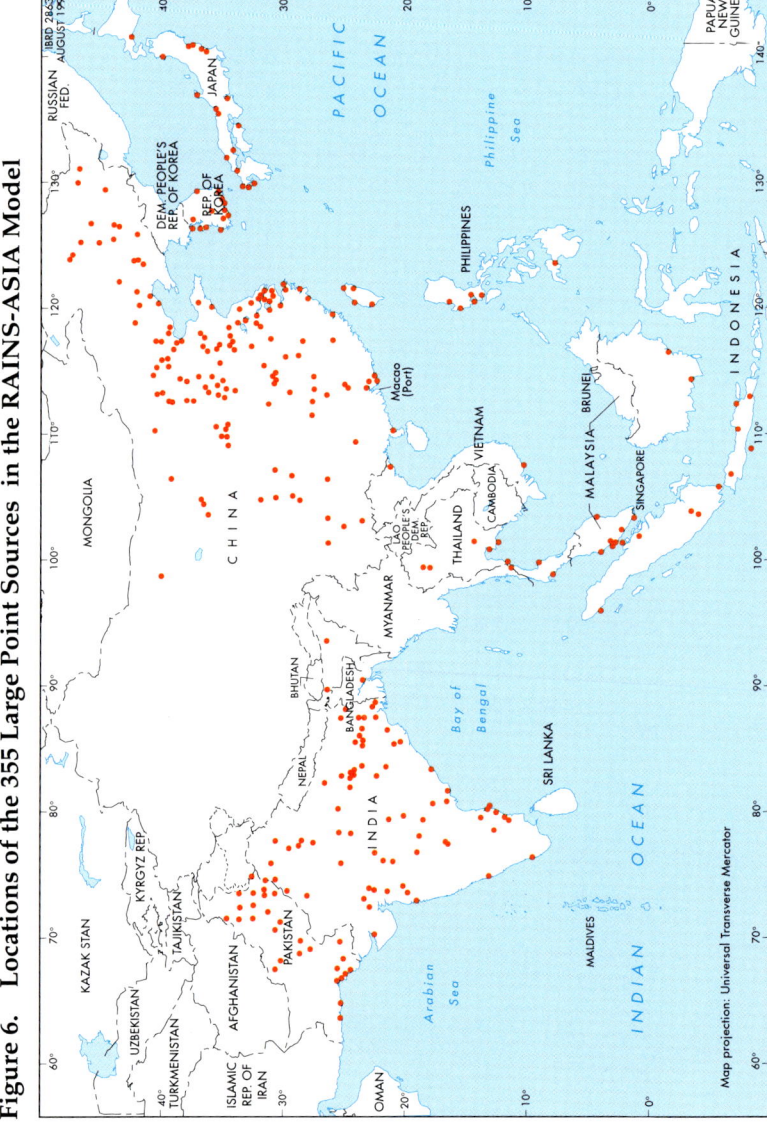

Map projection: Universal Transverse Mercator

Source: Green and others (1995).

Principal Results

Total emissions of sulfur dioxide for the base-case, no-control scenario, for each country are shown in table 3. Figure 7 shows the region's annual sulfur dioxide emissions in 1990.

A detailed analysis of scenario results, including calculations of emissions and control costs for a number of different scenarios, can be found in chapter 4.

Table 3.   Total Emissions of Sulfur Dioxide and the Average Annual Growth Rate of Emissions under the Base-Case, No-Control Scenario

Economy	Sulfur dioxide emission (kilotons SO_2)				Average annual growth rate, 1990–2020 (percent)
	1990	*2000*	*2010*	*2020*	
Bangladesh	118	165	330	525	5.1
Bhutan	2	5	7	12	7.0
Brunei	6	8	13	18	3.6
Cambodia	22	40	75	147	6.5
China	21,908	34,328	47,840	60,688	3.5
Hong Kong, China	140	216	290	378	3.4
India	4,472	6,594	10,931	18,549	4.9
Indonesia	630	1,085	1,868	3,162	5.5
Japan	835	997	1,048	1,120	1.0
Korea, Dem. Rep.	343	586	878	1,345	4.7
Korea, Rep. of	1,640	2,802	4,033	5,537	4.1
Lao PDR	3	5	8	12	4.3
Malaysia	206	242	342	410	2.3
Mongolia	78	95	124	168	2.6
Myanmar	18	25	32	40	2.7
Nepal	122	156	194	247	2.4
Pakistan	614	1,553	3,684	7,527	8.7
Philippines	391	627	1,071	2,037	5.7
Sea lanes	243	310	397	512	2.5
Singapore	191	358	653	1,033	5.8
Sri Lanka	42	132	171	239	6.0
Taiwan, China	500	765	1,086	1,478	3.7
Thailand	1,038	1,901	3,277	4,638	5.1
Vietnam	113	166	333	655	6.0
Total	33,675	53,161	78,685	110,478	4.0

Note: The model considers international sea (shipping) lanes as a separate region.
Source: RAINS-ASIA model, version 7.01.

Figure 7. Annual Sulfur Dioxide Emissions in Asia, 1990

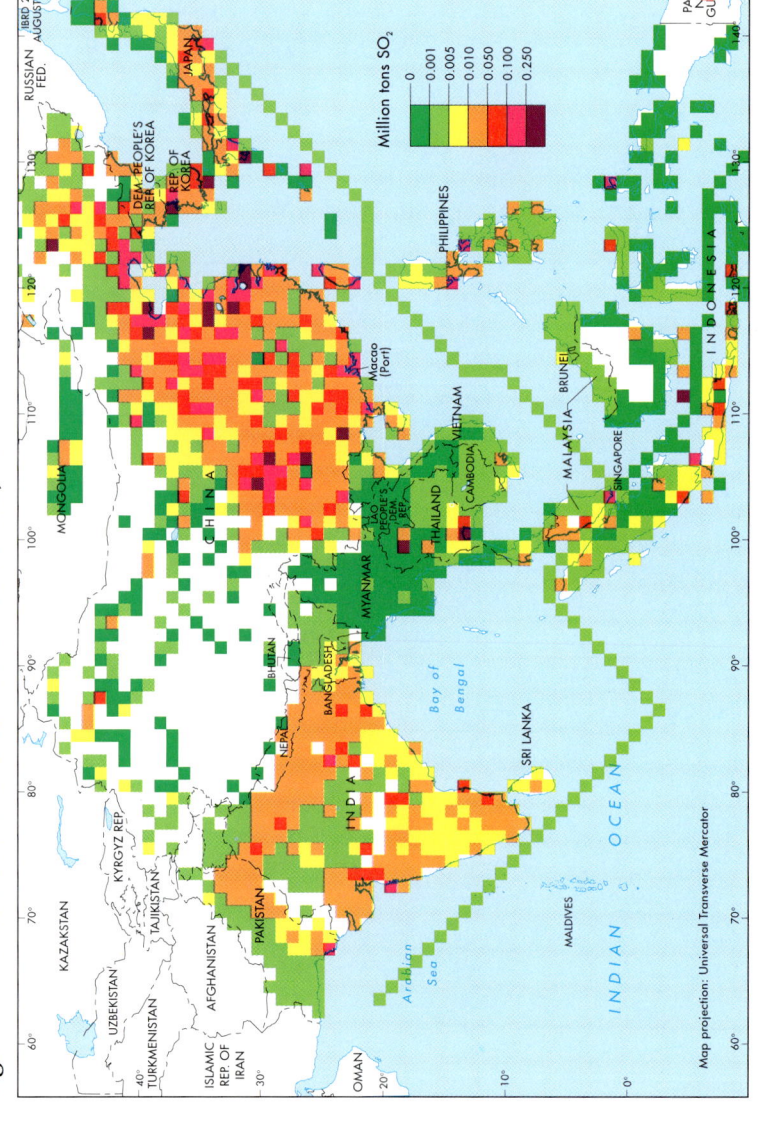

Source: Carmichael and Arndt (1995).

28

Deposition and Critical Loads Assessment (DEP) Module

The DEP module consists of two major components: the ATMOS submodule, which calculates atmospheric transport and deposition patterns for sulfur dioxide and sulfate, and the IMPACT submodule, which estimates the environmental effects of acid deposition to ecosystems across Asia. The function and structure of each of these submodules are described separately in the following sections.

Atmospheric Transport and Deposition (ATMOS) Submodule

The ATMOS submodule analyzes long-range transport and deposition of sulfur in Asia. The module calculates the sulfur deposition levels and patterns that result from various energy and emissions scenarios generated by the RESGEN and ENEM modules. ATMOS combines information on the location and levels of emissions from the other modules with meteorological, chemical, and physical data to calculate the resulting sulfur deposition patterns.

ATMOS follows individual parcels of air from a particular source throughout their trajectories—from emission, through atmospheric transport and chemical transformation, to deposition. The module incorporates an atmospheric source–receptor relationship, calculating sulfur transport and deposition, across all of Asia on a 1° by 1° grid and includes three separate horizontal layers (surface, boundary, and upper). ATMOS also takes into consideration variations in emission height (that is, it makes adjustments for emissions from tall stacks).

The submodule can be run for an entire year for each identified source, calculating the total annual deposition attributable to that source. Similarly, when run for all sources and areas, the model calculates the total annual deposition over the entire model region. Thus, ATMOS can be used to answer questions such as these:

- How do changes in energy consumption and emissions from a specific area or a single large point source affect levels of acid deposition in other areas?
- What sources or areas contribute to sulfur deposition in a given region?

Structure

The ATMOS submodule uses input data on emission rates, levels, and source locations (from ENEM). It also incorporates meteorological data (including winds, temperature, and precipitation rates) available from international organizations or national sources. Data on emissions supplied by the ENEM module comprise both anthropogenic and natural sources, including large point sources, area emissions (subdivided by industrial, domestic, and transportation categories), shipping activities (including regional shipping lanes and in-port activities), and active volcanoes.

The model uses meteorological data from the U.S. National Oceanic and Atmospheric Administration for approximately 200 stations in the region and precipitation data from the National Center for Atmospheric Research and monitoring sites in Asia. The model provides annual average (wet + dry) sulfate deposition values and monthly average sulfur dioxide concentration values for each 1° by 1° grid cell. Concentration and deposition values are calculated separately for large point and area sources. For dispersed-area sources, the results are aggregated, showing each region's contribution to deposition in a particular grid cell. Emissions from each large point source are calculated individually and show the contribution of each large point source to each grid cell. The module takes into account intra-annual variability while estimating the annual average deposition of sulfur.

The model allows the user to assess the spread of pollution from an individual source or region and identify the emission sources that contribute to sulfur deposition at a particular site. Table 4 shows the results of an analysis of the sources of sulfur deposition in Chongqing, China.

Principal Results

Figure 8 shows total annual sulfur deposition in Asia for 1990, including contributions from all anthropogenic sources in the model (large point and area sources) and volcanoes. The distribution pattern of sulfur deposition closely follows that of sulfur emissions (shown in figure 8).

Many areas with high emissions levels (for example, eastern and southern China, Korea, northern Thailand, and eastern India) show high levels of sulfur deposition. Annual precipitation levels also affect deposition patterns, as can be seen in areas with high precipitation such as northern India, Nepal, southeastern China, and Southeast Asia. High levels of sulfur deposition are also noted in major shipping lanes. In some areas below 20° north latitude, ship traffic accounts for 10 percent or more of annual sulfur deposition.

Model results for shorter periods of time indicate a strong seasonal variation, with large differences observed between the December–February and June–August periods. The effects of rainy seasons are reflected in the deposition patterns: more than 30 percent of the total wet deposition over the Japan Sea and Indonesia occurs during the three winter months (the winter monsoon period). Conversely, more than 30 percent of the total wet deposition in Southeast Asia and large parts of

Table 4.   Emission Sources Contributing to Sulfur Deposition in Chongqing, China

Region	Contribution of sulfur (milligrams per square meter per year)
Area sources	10,634
Hebei-Anhui-Henah	11
Shaanxi-Gansu	12
Hubei	34
Hunan	13
Guangxi	19
Sichuan	2,231
Chongqing	6,922
Guizhou	388
Guiyang	173
Yunnan	173
Large point sources	658
Chongqing, LPS4	550
Guizhou, LPS11	44
Sichuan, LPS56	64
Total	11,292

Note: Sources contributing less than 10 milligrams per square meter per year are not included.
Source: RAINS-ASIA model, version 7.01.

the Indian subcontinent occurs between June and August, the monsoon season in these areas.

   Another notable result from Phase I of the project has been the initiation of a passive sampler network at the forty-three sites in eleven countries described earlier (see figure 3 and table 2). Sample analysis began in early 1994, and monthly and annual mean concentration data are now available for all sites. Although no direct comparisons can yet be made between the network results (for 1994) and model-calculated values (for 1990), the relative spatial distribution of emissions (that is, areas of high and low concentrations) between modeled and measured values agree fairly well.

Environmental Impact and Critical Loads (IMPACT) Submodule

The IMPACT submodule assesses the sensitivity of various ecosystems (their "critical loads"; see box 4) to acid deposition and compares this information to the deposition data generated by the ATMOS module. This

Figure 8. Sulfur Deposition in Asia, 1990

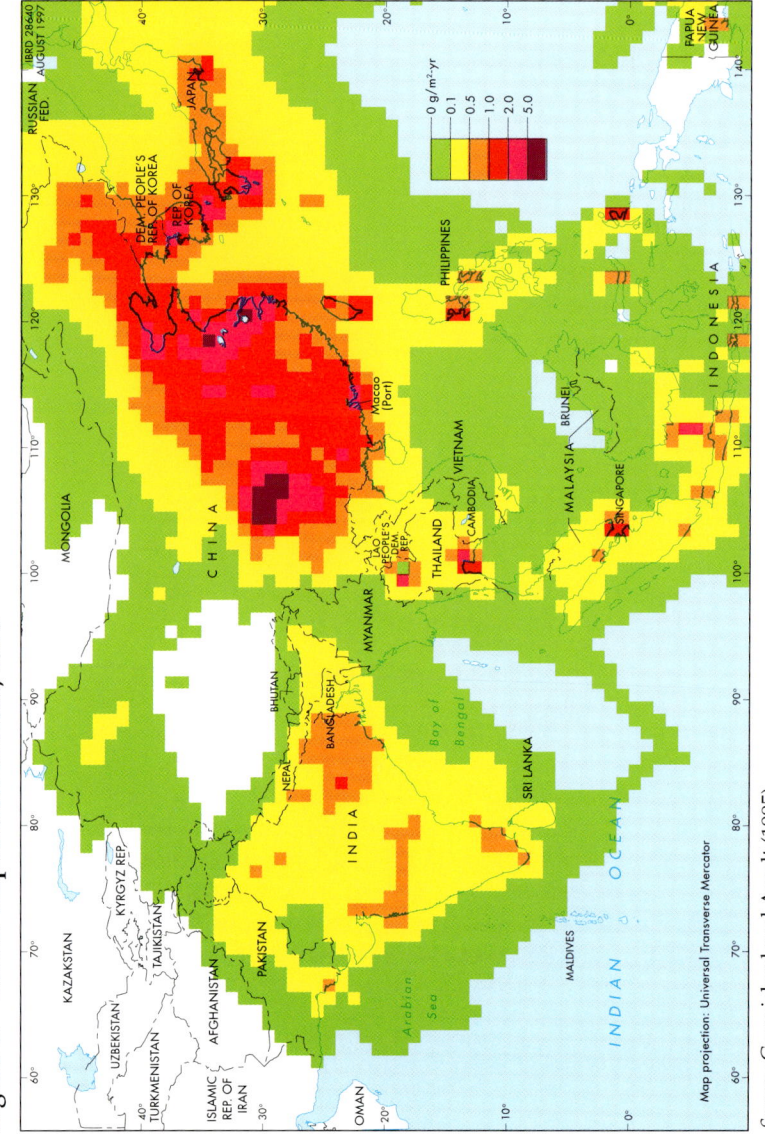

Source: Carmichael and Arndt (1995).

Box 4.   What Is a Critical Load?

A critical load of an ecosystem is a no-effect level for a pollutant (that is, the level of a substance—acid deposition, for example—that does not cause long-term damage to an ecosystem). Areas that have a limited natural capacity to absorb or neutralize acid rain have a low critical load. Ecosystems that are better able to buffer acidity (through different soil chemistry, biological tolerances, or other factors) have a correspondingly higher critical load. Assessing the natural capacity of ecosystems to withstand current and projected levels of pollution is a method of measuring ecosystem health and can serve as a way to assess the environmental benefits of emissions reductions.

---

process identifies regional ecological sensitivity and indicates which areas are at greatest risk of damage (for example, growth reduction, yield loss, or changes in biodiversity) from present or projected levels of sulfur deposition.

By estimating critical loads for various regions and ecosystems and comparing these natural sensitivities to deposition levels, the IMPACT submodule allows users to assess the environmental effects of different energy and emissions scenarios and answer the following questions:

- What regions and ecosystems are most sensitive to acid deposition?
- Which ecosystems are damaged, and to what extent, in a particular energy scenario?
- What are the environmental aspects of a particular emissions control strategy?

Structure

Two complementary methods are used to estimate critical loads for a variety of ecosystems: the definition of relative sensitivity classes and the steady-state mass balance method. The objective of applying two methodologies is to consider a large number of biogeochemical factors to determine the sensitivity of ecosystems, assess the reliability of each method by comparing the broad geographic distribution of results from each method, and extend the geographic scope of the assessment of ecosystem sensitivity to include areas in which data are insufficient to calculate critical load values directly. Each of these two methods is described in greater detail below.

The relative-sensitivity method uses information on climatic factors, geology, soil characteristics, vegetation type, and land use. These factors are categorized, weighted, and combined to define relative classes of ecosystem sensitivity to acid deposition. This method has been adapted from previous European applications of the relative-sensitivity method.

The following factors are considered in determining relative ecosystem sensitivity:

- Climatic factors such as annual average rainfall and temperature, runoff rates, and deposition of base cations that can neutralize acidity
- Soil chemistry and mineralogy, including soil pH, texture, geology, and rates of nutrient uptake and weathering
- Types of vegetation cover and land use.

The steady-state mass balance method determines the maximum level of a substance (sulfur-based acidity in the present model) that will not damage an ecosystem over the long term. In Phase I, critical loads have been calculated for thirty-one ecosystems in the region (see box 5). The maps showing critical loads and areas of excess sulfur deposition can be broken down by ecosystem type, including agriculture, rice paddies (an economically important crop), nonagricultural areas, or all ecosystems.

The steady-state mass balance method was used to calculate critical loads in the recent effects-based UN ECE protocol on sulfur emissions reductions (UN ECE 1994). This method has also been applied on a national basis in the RAINS-ASIA IMPACT submodule. The steady-state mass balance method was used on a site-specific basis in China, whereas in Japan, many versions of the steady-state mass balance model were implemented and tested on a regional scale. Because there were sufficient data to compute critical loads by the steady-state mass balance method, the relative-sensitivity method was used primarily to evaluate the robustness of the critical load distributions.

Several simplifying factors have been incorporated into the present version of the IMPACT submodule. Only the indirect environmental effects of sulfur deposition (that is, acidification) have been considered in

---

Box 5. Ecosystems Considered in the RAINS Model

Polar or rock desert	Interrupted temperate	Irrigated other farm-
Tundra	woods	land
Cool semidesert/	Dry or highland woods	Coastal wetland, cold
scrub	Mediterranean woodland	Coastal wetland,
Montane cold scrub	Interrupted tropical	mangrove
or grass	woods	Coastal wetland and
Cool scrub or	Subtropical dry forest	hinterland
grassland	Subtropical wet forest	Hot scrub or grassland
Main taiga	Tropical dry forest	Succulents and thorn
Southern taiga	or dry woods	Semiarid desert
Coniferous forest	Tropical wet forest	Nonpolar rocky
Mixed forest	Tropical savanna	vegetation
Temperate broadleaf	General farmland	Sand desert
forest	Irrigated paddy	Semidesert

Figure 9.   Critical Loads for Acidity

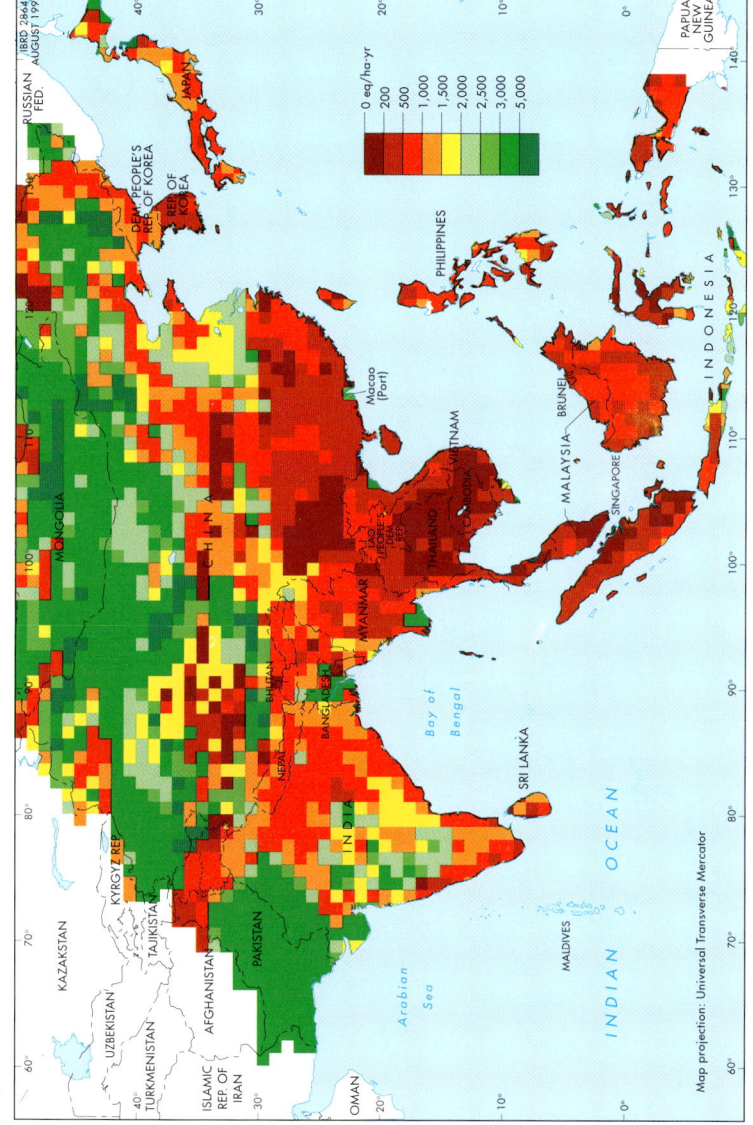

Source: RAINS–ASIA 7.02.

35

Figure 10. Excess Sulfur Deposition above Critical Loads, 1990

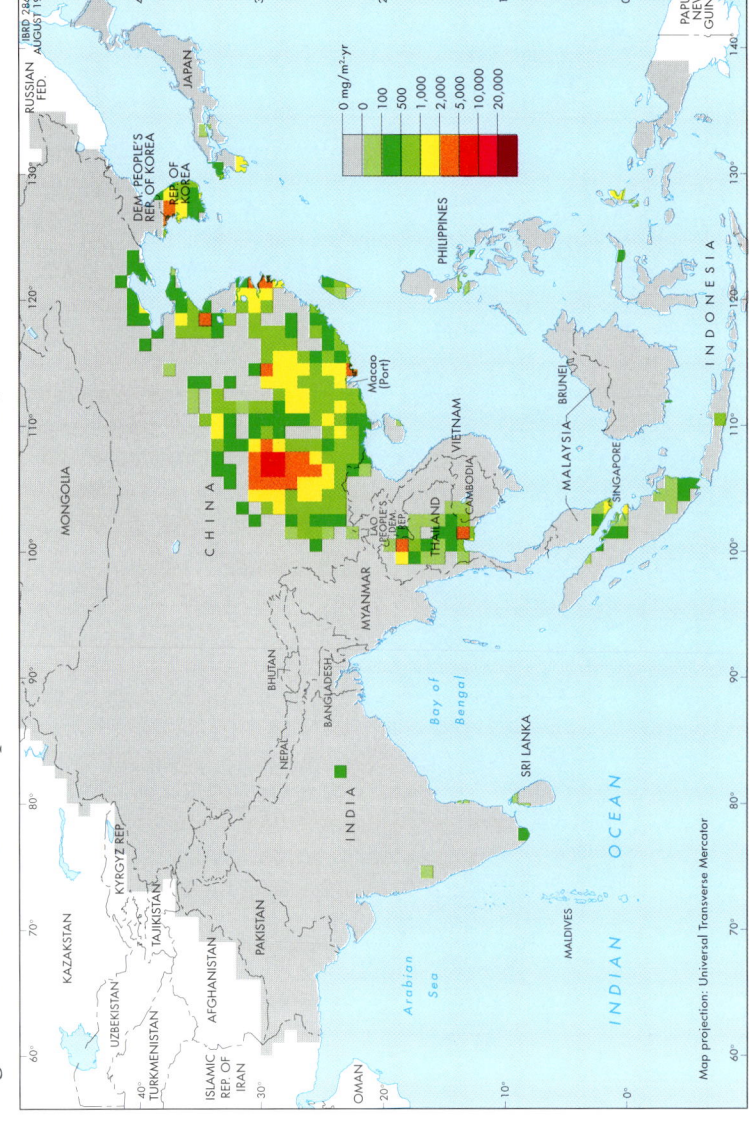

Source: Hettelingh and others (1995).

the present method; other direct effects such as damage to vegetation from elevated sulfur dioxide air concentrations are not yet considered in the model. As is the case with all other parts of the RAINS-ASIA model, the IMPACT submodule considers only the role of sulfur in acidification. Although nitrogen also plays an important role in the acidification process, for the first phase of the project, emphasis was placed on the effects of large-scale sulfur dioxide emissions and deposition.

The IMPACT submodule is not a dose-response module. Its results only tell the user where sulfur deposition exceeds carrying capacity. The model does not estimate the effect of that excess, and hence the extent of ecosystem damage and economic costs cannot be quantified.

Principal Results

A comparison of the results of the steady-state mass balance model and the relative-sensitivity method reveals overall compatibility between the two approaches in determining the distribution of sensitive areas. In most areas in which the relative-sensitivity method indicated the likelihood of sensitive ecosystems (based on geologic, land use, and climatic data), the steady-state mass balance model also calculated relatively low critical loads. Thus, the reliability of the initial results is improved, although large variations in the availability of input data remain.

Figure 9 shows the map of critical loads for acidity in Asia. Because a single grid cell can contain numerous combinations of vegetation, soil type, and other factors that influence the critical load, the map shows critical load values that protect 75 percent of all ecosystems in that grid cell. Figure 10 shows the areas in which sulfur deposition in 1990 exceeded the critical load, thus endangering these ecosystems to damage by acidification.

4
Results from the First Phase

A number of preliminary analyses have already been carried out using the RAINS-ASIA model. The model provides the first comprehensive and integrated analysis of the environmental consequences of continued uncontrolled growth of energy consumption and sulfur emissions in Asia. The reference scenario described earlier has been used as a foundation for developing a number of other scenarios that are described in the following sections. These scenarios investigate the effectiveness of different pollution-control strategies. A number of control options were developed, and the resulting costs and environmental effects, including improvements in sulfur emissions, deposition, and overall effects, were quantified.

Base-Case (Reference) Scenario

The base-case scenario assumes continuation of present economic and environmental trends with no additional measures to reduce sulfur (see chapter 3). In this scenario, total energy demand would increase at an average rate of 4 percent annually from 1990 to 2020 (figure 6), resulting in a tripling of energy use during this thirty-year period. Coal combustion would continue to be the primary source of sulfur emissions, accounting for about 75 percent of total emissions as shown in figure 11.

Total emissions would more than triple, from 33.6 million tons in 1990 to more than 110 million tons by 2020. Emissions growth rates would vary widely among countries, as shown in table 3. Emissions levels in Japan are projected to rise only 30 percent under the base-case scenario, whereas 400 to 500 percent increases would be experienced in countries such as India, Indonesia, the Philippines, and Thailand. Emissions from power plants are projected to grow most rapidly as a result of the increased use of coal to generate electricity (see figure 12). Another notable trend is the increased contribution of large point sources to the overall emissions situation. In 1990, the large point sources considered

Figure 11.   Sulfur Dioxide Emissions, by Fuel, under the Reference Scenario

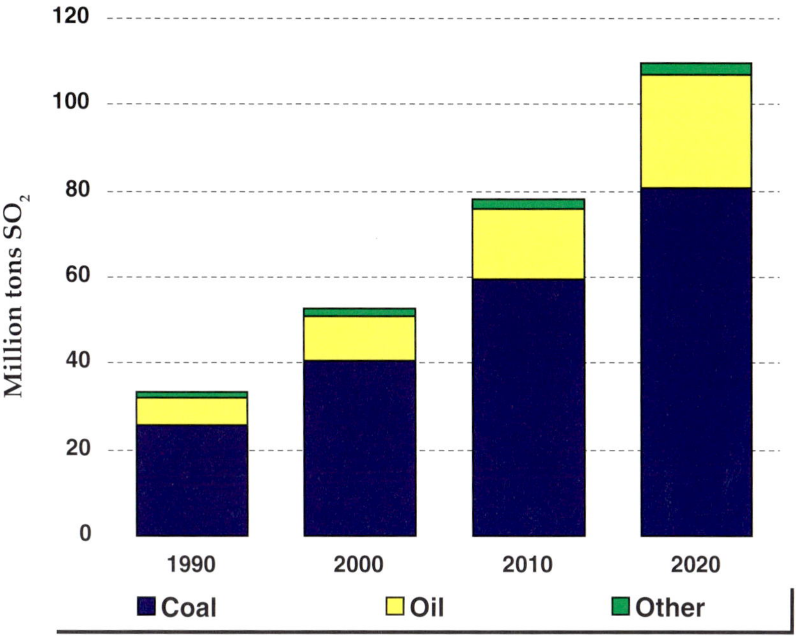

Source: Amann and Cofala (1995).

in the model were estimated to contribute approximately 16 percent of the total sulfur dioxide emissions. In the reference scenario, by 2020 the share of emissions originating from these sources would increase to 25 percent. The sulfur deposition patterns from these increased emissions are shown in figure 13.

Large parts of eastern China and most of India would receive between 2 and 5 grams of sulfur per square meter per year. Many industrialized areas of Indonesia, Malaysia, the Philippines, and Thailand would experience sulfur deposition levels of 5 to 10 grams per square meter per year, whereas local hot spots in some industrialized areas of China would receive about 18 grams of sulfur per square meter every year. As expected, the enormous growth in emissions and deposition under the reference scenario would lead to significantly higher levels of ecosystem damage as well. Figure 14 displays the excess sulfur deposition (that is, deposition that exceeds the critical load) in 2020. The map shows that large portions of northern and eastern India, southern and eastern China, parts of northern and central Thailand, and much of the Korean peninsula will experience sulfur deposition levels that exceed the ecosystem critical load in those areas.

Figure 12.   Sulfur Dioxide Emissions, by Sector, under the Reference Scenario

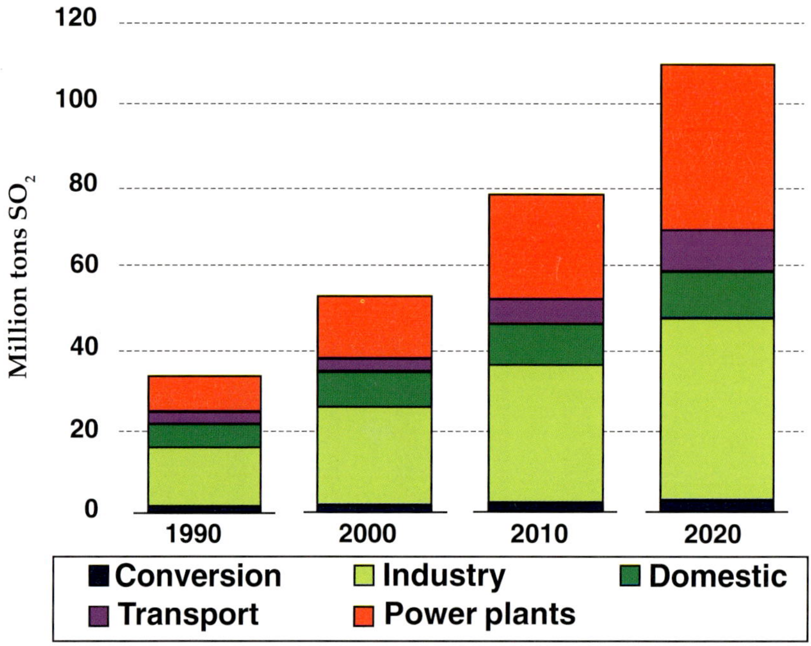

Source: Amann and Cofala (1995).

Because of the wide variety of ecosystem types and climates encountered in Asia, the complexities of the biogeochemical processes involved in the acidification process, and the lack of monitoring data, it is not presently possible to precisely quantify the environmental damage resulting from excess sulfur deposition. That some areas could receive ten times or more sulfur than these ecosystems can tolerate, however, indicates the potential for widespread ecosystem injury.

Similarly, high levels of sulfur dioxide concentrations are also predicted under the no-control policies assumed in the reference scenario. The grid-based model calculations show large areas, particularly in China and India, near or exceeding WHO guidelines for ambient air quality (figure 15). Urban air quality monitoring data tend to indicate that sulfur dioxide levels in cities are considerably higher than the grid-average values calculated by the model. Additional investigation of air quality issues is envisioned as an important part of further development of the model.

Basic Control Technology (BCT)

In view of the level of financial resources required to implement sophisticated emissions control techniques on a broad scale, it is necessary to con-

Figure 13.   Sulfur Deposition in 2020 under the Reference Scenario

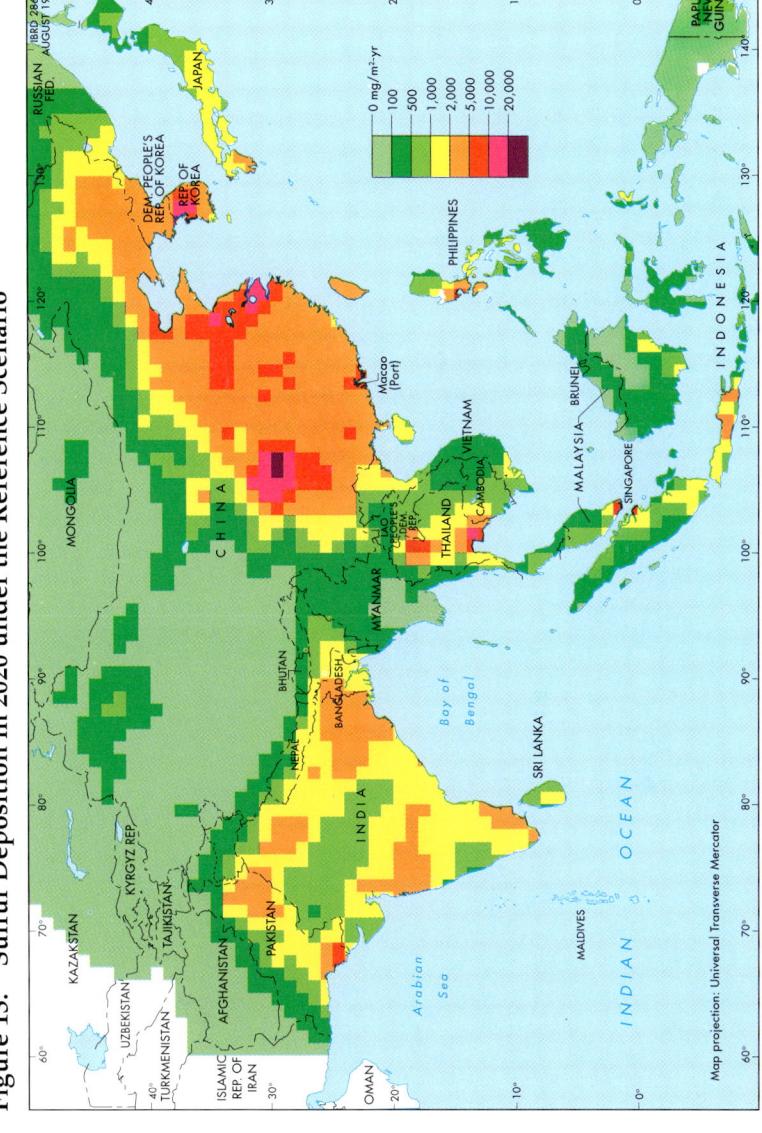

Source: Carmichael and Arndt (1995).

41

Figure 14. Excess Sulfur Deposition above Critical Loads in 2020 under the Reference Scenario

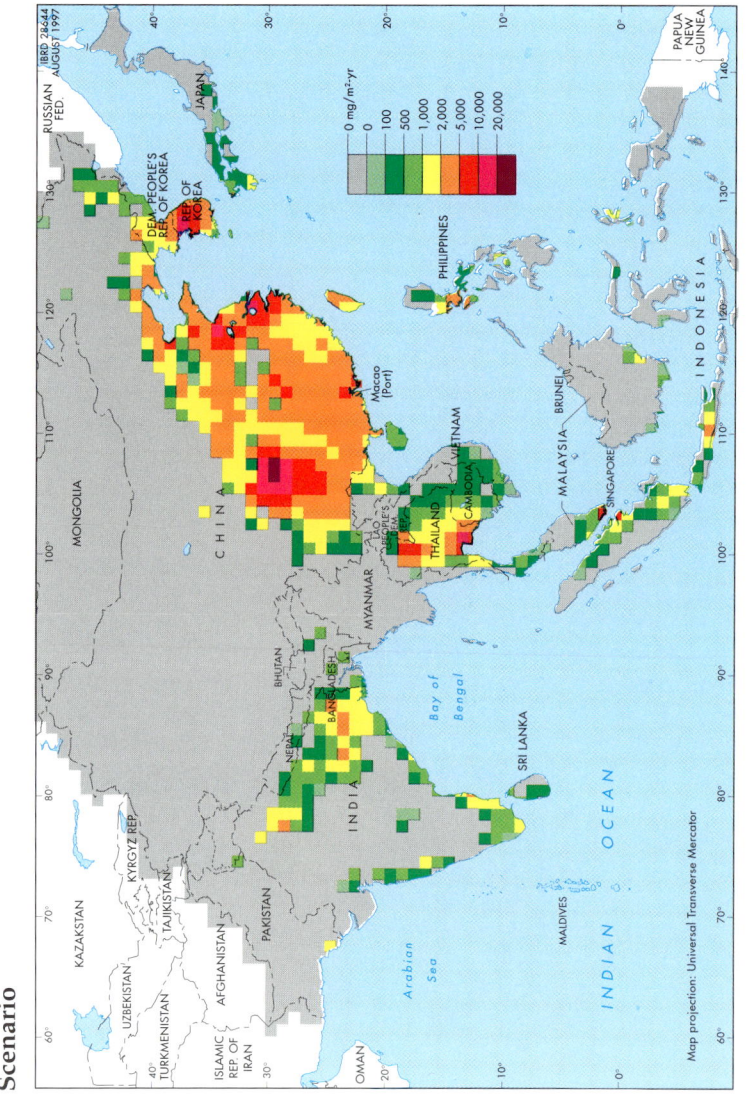

Source: Amann and Cofala (1995).

Figure 15. Ambient Levels of Sulfur Dioxide Concentration in 2020 under the Reference Scenario

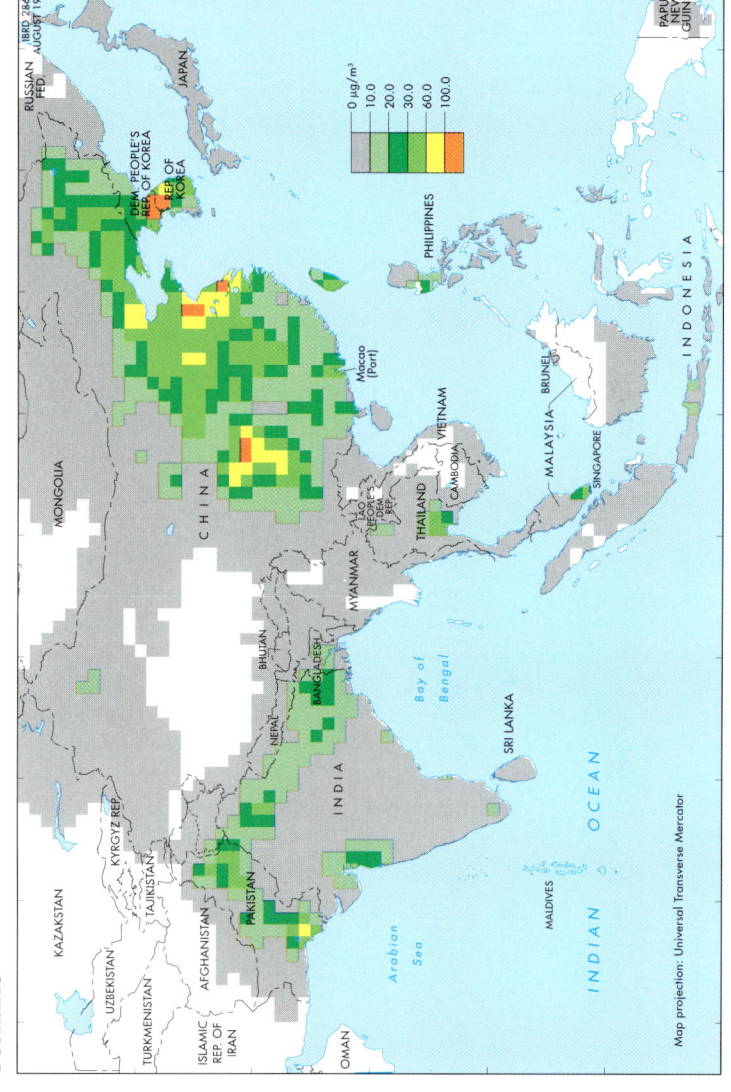

Source: Amann and Cofala (1995).

43

sider less advanced, and often less costly, solutions. The BCT scenario considers more basic, domestically available control technologies, such as limestone-injection procedures for power plants, which remove only about 50 percent of the sulfur in fuel but also require lower capital investments than other, more technologically advanced, options (see tables 5 and 6).

As expected with the use of less-effective emissions control technology, emissions levels in the three countries taken into account (China, India, and Pakistan) are considerably higher than those projected for other, more stringent, scenarios, such as ACT. Nevertheless, total control costs are roughly the same for the two scenarios, primarily because the RAINS-ASIA model calculates total life-cycle costs for emissions control tech-

Table 5.   Emissions Levels of Sulfur Dioxide and Control Costs per Country for the BAT, ACT, and BCT Scenarios

Economy	SO_2 emissions (thousand tons)			Control costs (million U.S. dollars per year)		
	BCT	BAT	ACT	BCT	BAT	ACT
Bangladesh	258	165	258	228	475	228
Bhutan	4	3	4	9	7	9
Brunei	17	15	17	2	15	2
Cambodia	69	22	69	123	487	123
China	38,124	6,672	29,932	12,712	34,230	11,975
Hong Kong, China	68	24	68	255	574	255
India	13,054	5,906	10,522	6,213	17,055	6,328
Indonesia	785	438	785	2,255	6,121	2,255
Japan	1,047	393	1,047	3,458	6,132	3,458
Korea, Dem. Rep.	7,075	75	7,075	1,089	3,087	1,089
Korea, Rep. of	1,469	552	1,469	3,214	3,769	3,214
Lao PDR	7	5	7	6	9	6
Malaysia	246	66	246	163	843	163
Mongolia	81	13	81	56	138	56
Myanmar	37	32	37	5	32	5
Nepal	230	218	230	12	53	12
Pakistan	3,609	606	1,907	3,703	4,333	3,095
Philippines	440	146	440	1,063	1,201	1,063
Sea lanes	307	102	307	222	445	222
Singapore	221	65	221	635	860	635
Sri Lanka	53	37	53	173	222	173
Taiwan, China	827	245	827	1,249	2,999	1,249
Thailand	813	336	813	2,916	6,485	2,916
Vietnam	345	183	345	338	853	338
Total	69,186	16,319	56,760	40,099	90,425	38,869

Note: The model considers international sea (shipping) lanes as separate regions.
Source: Amann and Cofala (1995).

Table 6.   Emissions and Control Costs for the Base-Case Energy Pathway Compared with the Energy-Efficiency Pathway, for Three Different Scenarios

	Emissions control scenario			
Energy pathway	No further control	BCT	ACT	BAT
Emissions (million tons SO$_2$)				
Base case	110.5	62.8	50.4	16.3
Efficiency	80.1	47.1	39.1	12.4
Costs (billions U.S. dollars per year)				
Base case	3.9	40.1	38.8	90.4
Efficiency	2.0	26.9	25.5	65.6

Source: Amann and Cofala (1995).

nologies. Although the controls implemented in the BCT scenario require less up-front capital investment, they have higher operating costs, most notably for handling large amounts of waste material produced.

The excess sulfur deposition resulting from the BCT scenario is shown in figure 16. Growth in emissions, particularly from large point sources, leads to large areas receiving excess sulfur deposition in the range of 2 to 5 grams of sulfur per square meter, with Sichuan and Shanghai provinces receiving 10 grams or more of excess sulfur per square meter. These results indicate that, in the long term, emissions control strategies which rely on technologies that remove only moderate amounts of sulfur will not be able to protect important agricultural areas from serious excess deposition.

Local Advanced Control Technology (LACT)

Although the present version of the model cannot optimize control strategies, the model development team has also conducted some preliminary investigations to maximize the cost-effectiveness of emissions-reduction strategies. The team found that overall control costs could be reduced while maintaining the same levels of environmental protection by targeting emissions control measures to sources in environmentally sensitive regions.

Under the LACT scenario, no additional controls are implemented in relatively low-income countries (such as Bangladesh, Cambodia, and Sri Lanka), whereas emissions from countries with a higher per capita income (such as Indonesia, Japan, Korea, and Thailand) are controlled. This scenario results in emissions levels roughly comparable to those under the BCT method, while costs are reduced by approximately one-third. Figure 17 depicts the excess sulfur deposition pattern under the LACT scenario.

Figure 16. Excess Sulfur Deposition above Critical Loads for the BCT Scenario in 2020

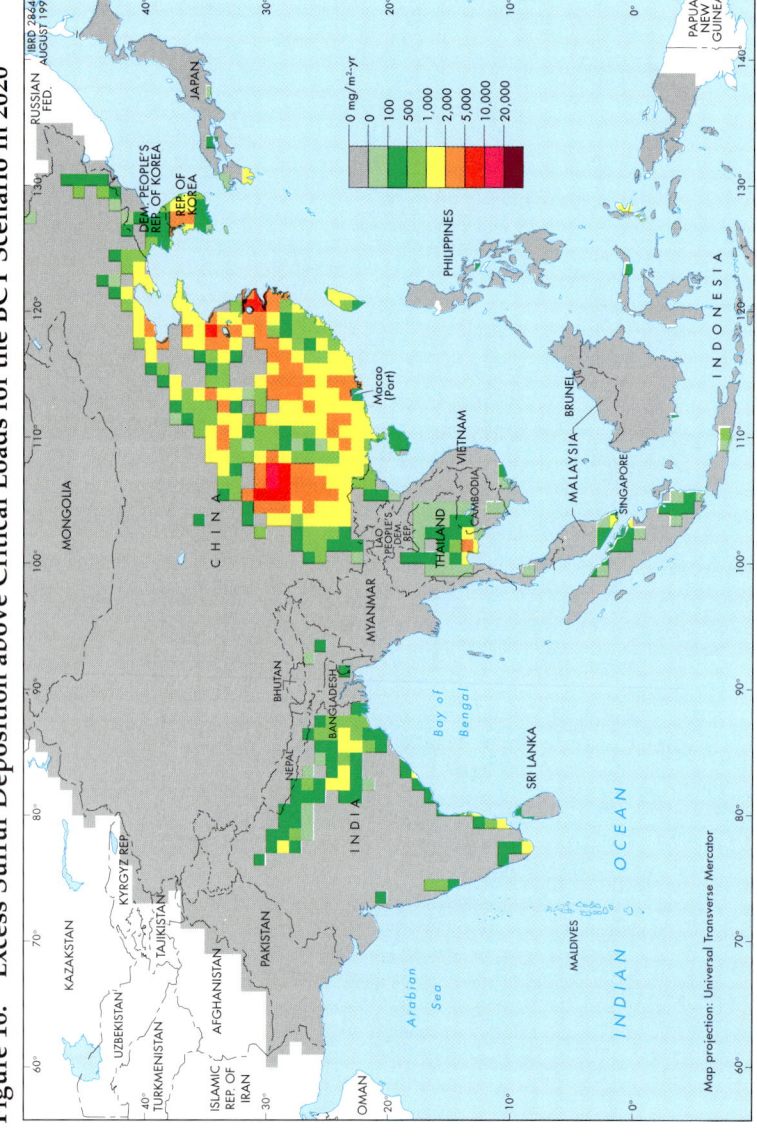

Source: Amann and Cofala (1995).

Figure 17. Excess Sulfur Deposition above Critical Loads for the LACT Scenario in 2020

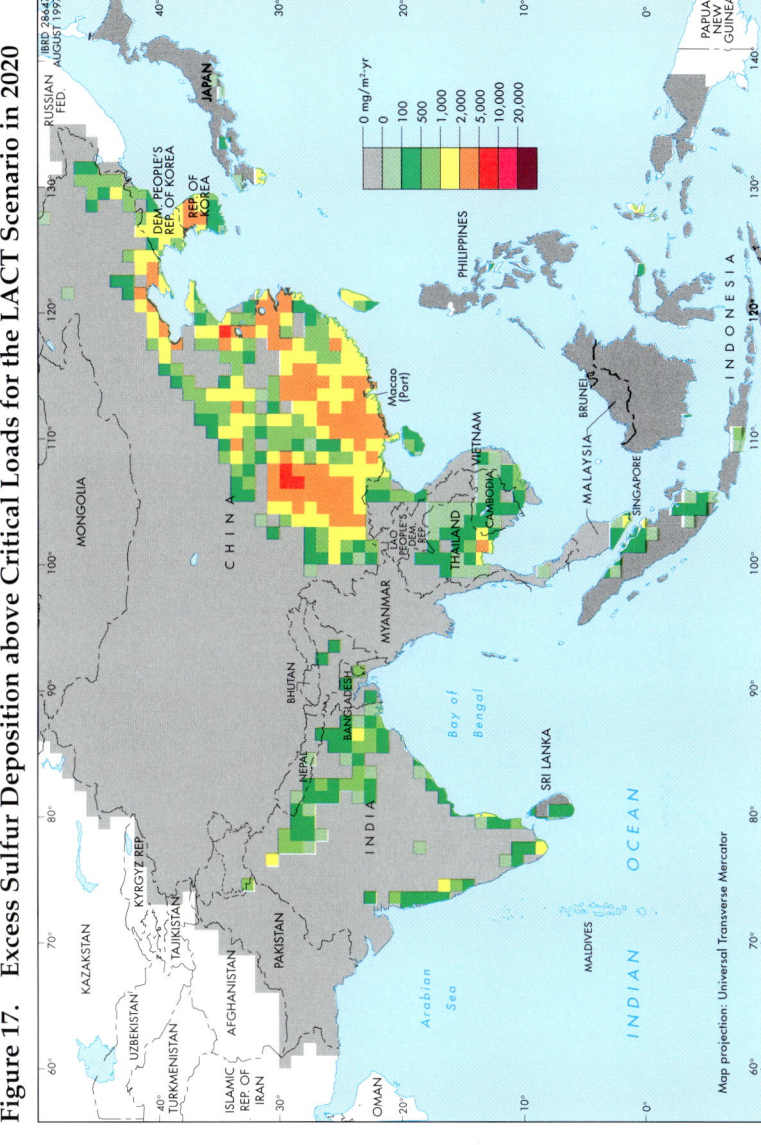

Source: Amann and Cofala (1995).

47

Best Available Technology (BAT)

In contrast to the base-case scenario, which assumes no new measures are taken to control emissions, the BAT scenario investigates the results of implementing state-of-the-art pollution control technologies in many sectors. Whereas the reference scenario is a worst-case analysis of the future situation, the BAT method can be seen as a best-case approach. In this scenario, wet flue-gas desulfurization technology is installed for all current and planned large point sources that burn coal or oil. For the residential, commercial, and transportation sectors, the use of low-sulfur fuels (coal and oil) is assumed.

The RAINS-ASIA model shows that drastic reductions in sulfur dioxide emissions can be achieved through the widespread introduction of advanced control technologies. Under the BAT scenario, sulfur dioxide emissions decrease by more than 50 percent in thirty years, from a 1990 level of 33.6 million tons to 16.3 million tons by 2020. The effect on deposition levels and critical loads exceedances is also remarkable. As shown in figure 18, nearly all areas will reach sustainable levels of sulfur deposition (that is, levels that avoid ecosystem damage). Some remaining problem areas still exist, however, including those around the border between Hunan and Jiangxi provinces in China, an area with robust industrial activity situated in a region of sensitive ecosystems. Additional local areas with high exceedances occur in India, Korea, and Thailand.

Costs associated with the extensive implementation of stringent control methods are just as striking as the environmental improvements. It is estimated that in 2020, the cost of carrying out the BAT strategy across Asia will be approximately US$90 billion per year, or about 0.6 percent of the region's gross domestic product. Individual countries' burdens would vary with level of economic development and reliance on heavily polluting fossil fuels, ranging from 0.05 of GDP in Myanmar and 0.06 percent for Japan to 1.7 percent for China.

Advanced Emission Control Technology (ACT)

Because the BAT scenario imposes significant financial burdens for many developing countries in the region, other scenarios were developed to try to define more cost-effective measures to reduce sulfur emissions. With the objective of protecting sensitive ecosystems, it is possible to rank a variety of control techniques in terms of their cost-effectiveness (that is, how much sulfur is reduced for a certain amount of money) and implement only those required to achieve critical loads. Thus, the ACT scenario selects those control methods (described in box 6) that reduce sulfur emissions at the lowest cost.

Figure 18. Excess Sulfur Deposition above Critical Loads for the BAT Scenario in 2020

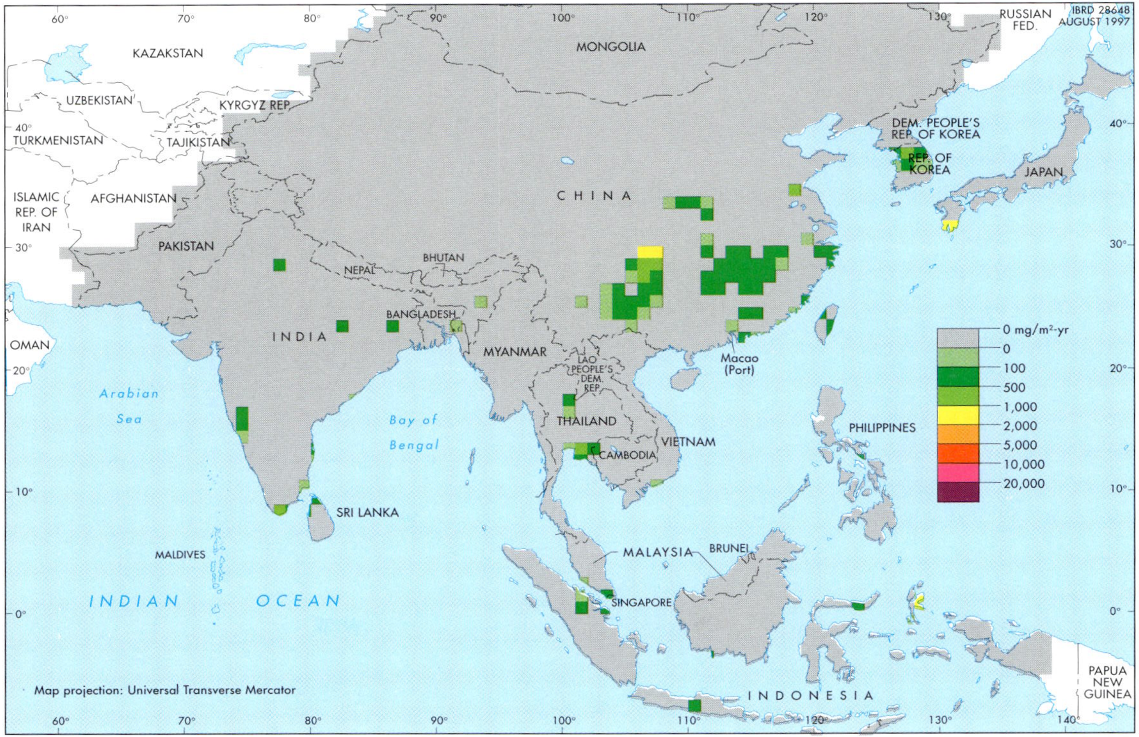

Source: Amann and Cofala (1995).

49

Box 6.   Measures Considered for Each Control Scenario

Base-case (reference) scenario
- Japan and Taiwan, China: Current and planned emissions control legislation is implemented.
- Other countries: No further legislation or policies to decrease sulfur emissions.

BCT
- China, India, and Pakistan: Domestic technologies with low capital requirements (for example, limestone injection) are implemented for all new coal-fired power stations. The domestic and transport sectors use entirely low-sulfur fuels.
- Other countries: Controls as in the ACT scenario.

LACT
- Low-income countries (such as Bangladesh and Cambodia): no controls implemented
- China, India, and Pakistan: Advanced emissions control measures implemented in certain regions to protect sensitive ecological areas
- Other countries: Controls as in ACT scenario

BAT
- Flue-gas desulfurization technologies (wet limestone scrubbing) are incorporated for all existing and future large power stations that burn coal or oil.
- Flue-gas desulfurization technologies (wet limestone scrubbing) are incorporated in all large industrial boilers.
- Domestic and transport sectors use low-sulfur fuels (coal, heavy fuel oil, and gas oil).

ACT
- Flue-gas desulfurization technologies (wet limestone scrubbing) are incorporated for all new power stations.
- Flue-gas desulfurization technologies (wet limestone scrubbing) are incorporated in all large industrial boilers in refineries.
- Low-sulfur fuels are used for industrial boilers (all liquid fuels and half of all coal consumption).
- Existing power stations, small industrial sources, and domestic and transport sectors use entirely low-sulfur fuels.
- Japan and Taiwan, China: Current and planned emissions control legislation is implemented.

The use of this method reduces not only the costs involved in emissions control but also the extent of emissions gains. Although sulfur dioxide emissions increase to more than 50 million tons by 2020, a 50 percent increase from 1990 levels, this increase is less than half of the future emissions level of 110 million tons calculated for the no-control reference scenario.

The ACT scenario entails considerably lower costs than the BAT strategy. Estimated costs total US$39 billion per year compared with the BAT's US$90 billion cost. National costs again vary between countries, although they are uniformly lower. The pan-Asian cost of implementing this scenario drops to 0.25 percent of GDP, a figure roughly equivalent to that of the recent European agreement to reduce sulfur emissions (0.21 percent GDP). The environmental results of the ACT scenario, in terms of excess sulfur deposition, are shown in figure 19.

A comparison of emission levels and control costs per country for the BAT, ACT, and BCT scenarios is shown in table 5.

Other Emissions Control Options

The preceding scenarios are based on the base-case energy pathway that forecasts a tripling of energy demand during the thirty-year period of the model analysis. These scenarios, however, do not consider other nontechnological methods of reducing emissions, such as promoting energy efficiency and using cleaner fuels. As shown in figure 1, such policies can have a dramatic effect on total energy demand.

A number of scenarios were reassessed in combination with the energy-efficiency pathway described earlier. Because these structural improvements in the energy system lead to lower fuel consumption, and thus lower sulfur dioxide emissions, the resulting control costs are lower than the comparable strategies that are based on base-case energy projections. Table 6 compares emission levels and control costs of three scenarios for both energy pathways.

Energy-efficiency measures are strong and cost-effective options for reducing emissions. These measures often produce secondary positive effects in addition to reducing emissions. Examples of such benefits include replacement of inefficient capital stock and a reduction in overall energy demand that in turn can improve trade balances.

Application of the Model

To incorporate the model in-country, it would be helpful if each participating country could establish a focus group for RAINS-ASIA. This group may take the form of a national steering committee composed of representatives from the ministries or agencies of energy, environment, planning, fuel (including petrochemicals), agriculture, and meteorology. Technical working groups focusing on energy and emissions, atmospheric transportation and deposition, and environmental impacts would come under the umbrella of this steering committee. It would be the responsibility of these working groups to validate and refine the model by

Figure 19. Excess Sulfur Deposition above Critical Loads for the ACT Scenario in 2020

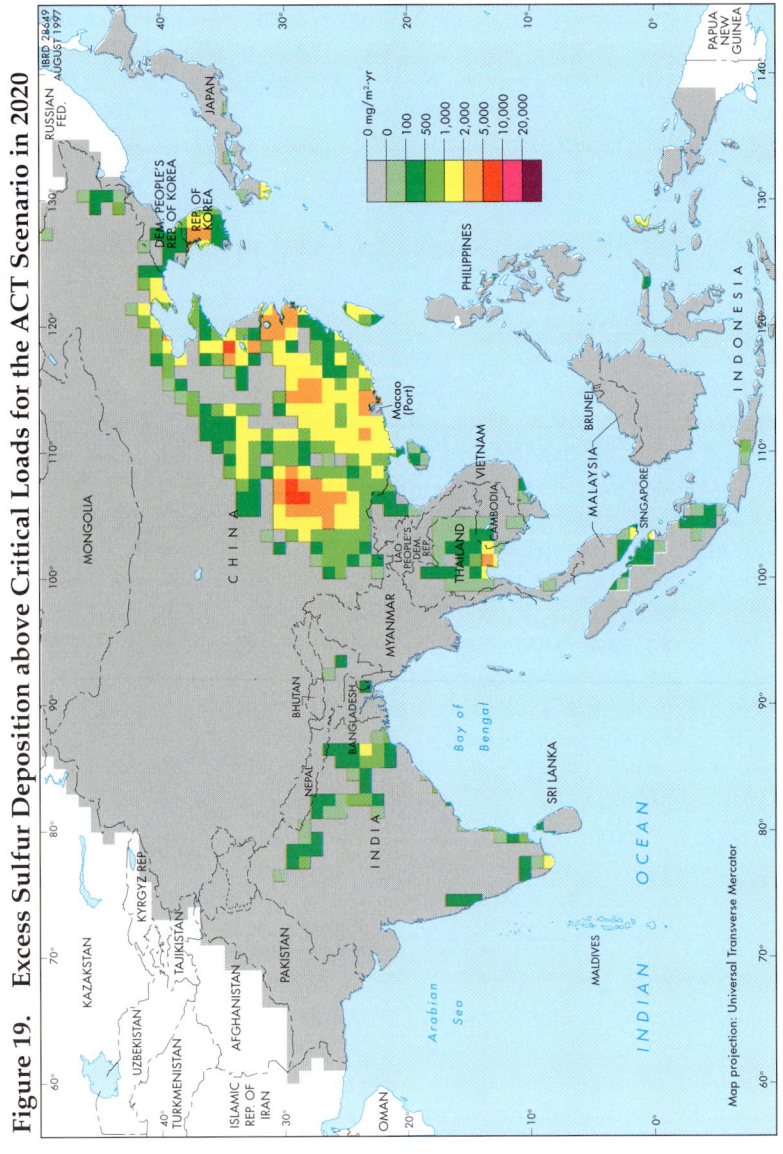

Map projection: Universal Transverse Mercator

Source: Amann and Cofala (1995).

examining conditions specific to the subregion or country and updating the data bases. Validation could be done through case studies for specific subregions or countries as a whole. The steering committee would provide broad guidance for these activities, explore applications of the model in country policy and planning activities, and provide guidance for future phases of the RAINS-ASIA program.

Technical working groups in different Asian countries could form a network and discuss issues formally and informally in national and international forums. This network would provide a framework to relate to other international scientists and policymakers working on the subject, thus furnishing a vehicle for effective cross-fertilization of ideas.

5

Conclusions and Future Work

Although many countries have recently started limiting sulfur dioxide emissions, which will help to contain the regional effect of such emissions, current trends in energy production and consumption in Asia indicate that acidification and air pollution problems are likely to worsen rapidly in the next thirty years. If 1990 trends continue, emissions of sulfur will more than triple throughout the region, and in many areas, sulfur deposition will grow to five to ten times the current levels. The incidence of harm to natural ecosystems, economically important crops, and human health will increase dramatically.

The RAINS-ASIA model has been designed as a forecasting tool to investigate the consequences of a variety of energy development scenarios for Asia. The model builds on the legacy of its European counterpart to provide a spatially detailed, comprehensive analysis of all stages of the acidification phenomenon: energy demand, supply, and production; emissions; atmospheric transport and deposition of acidifying compounds; and environmental effects of current and predicted levels of acid deposition.

RAINS-ASIA is a powerful tool for assessing the emissions, costs, and environmental consequences of a variety of future energy and environmental scenarios. In only a few years, enormous amounts of data have been collected and incorporated into the model to gauge the potential environmental effects of current and future sulfur dioxide emissions in Asia.

Preliminary analyses of future scenarios of energy use and sulfur pollution in Asia show that steps can be taken to avoid the most drastic consequences of continued, uncontrolled energy and emissions growth. Although the costs are high, the model can help realize maximum benefits from alternative measures. The costs of doing nothing are likely to be much greater in the long term.

A fundamental prerequisite for further refinement of the model is feedback from the model's end users: scientists, researchers, and policymakers in the regions. To this end, the widespread dissemination and use of the model is already under way through information and training workshops for in-country managers and policymakers.

The RAINS model provides an important opportunity to assess the regional acidification impacts and associated costs of various national and regional energy development strategies. Further involvement of in-country policymakers, managers, and scientists is needed to refine the model assumptions and input data; thoroughly test the model assumptions and input data; and analyze, disseminate, and implement model findings.

The RAINS-ASIA program is a blossoming effort to understand and deal with regional air pollution issues resulting from Asia's energy consumption. Although it has several limitations in its current stage of development, the model provides a useful tool for looking at future scenarios of energy growth and resulting environmental consequences. The program has built an international network of scientists, policymakers, multilateral funding institutions, and other donors with the motivation to help avoid future environmental problems. The World Bank and the Asian Development Bank hope to be catalysts in this process by providing analytical and financial support wherever appropriate and bringing to bear experiences from other countries.

Future Modifications of the Model

It is envisioned that the process of refining and updating the model will be a continuous one. Priorities for further developing the model are listed in this section.

Update the inventory in the Energy and Emissions Module and incorporate more Asia-specific data. To assess and compare various emissions control strategies more accurately, it is important to have complete and up-to-date information on technologies, costs, and applications relevant to the region. In the current version of the model, control costs and technology performance data are based on Western experiences and originate in Europe and North America. An Asia-specific database would include additional analysis of improvements in energy efficiency, low-cost technologies not normally applied in the West, and indigenous fuel resources (for example, Thai lignite and high-ash Indian coal).

Improve meteorology used in the transport and deposition module. The model needs to be run for multiple years of meteorological data to provide better estimates of deposition and ambient concentrations. Only by being used for multiple years of data can the model accurately assess interannual variability of air pollution and resulting deposition patterns.

Also, the scale of deposition requires refinement from a 1° by 1° calculation to finer grid sizes.

Include additional pollutants and ecosystems. To obtain a complete view of the entire acidification phenomenon and to better assess urban pollution problems, emissions of nitrogen oxides must be factored into the model. This expansion of the model will require more focus on the transportation sector. In addition, more regional information is needed on the potential effects of air pollutants on a variety of other receptors such as aquatic ecosystems, infrastructure (buildings and materials), and human health.

Expand critical load strategy and quantification of benefits. In addition to calculating environmental effects in terms of sulfur-based critical loads, an assessment of the additive and synergistic effects among a variety of pollutants would be helpful. An analysis of the relative valuation of current and future use of natural resources, as well as the direct and delayed effects on the quality of those resources (that is, damage assessment), would also be useful. Such damage assessment allows an evaluation of regional benefits of abatement policies aimed at sustainability. Dynamic modeling, to determine the temporal aspect of ecosystem acidification and damage, would be a main aspect of this strategy.

Develop optimization capabilities. The current model operates solely in a scenario analysis mode; that is, it calculates the results in emissions, deposition, and damage based on a set of energy assumptions. An optimization routine works essentially in reverse: it uses end points defined in terms of environmental targets (and economic constraints) and calculates the cost-optimal solution, in terms of emissions controls required, to reach these targets. The addition of such optimization capabilities to the model would improve its abilities as a decisionmaking tool.

Complete additional analyses. Additional work is needed to address issues such as the possible use of economic incentives to achieve emissions reductions and the estimation of the economic costs of acid deposition. Each portion of the model and its databases, from the assumptions contained in the energy and emissions scenarios, to atmospheric transport modeling, to the estimation of environmental effects, have an inherent range of uncertainty. A sensitivity analysis to assess and improve the overall reliability of the model would be helpful in interpreting the results.

Integrate RESGEN into RAINS. Given that the interface between RESGEN and RAINS is complex and inefficient, it would be necessary to develop a scenario-generating module as an integral part of RAINS-ASIA.

Evaluate and validate the model. Evaluation and validation are critical to improving the credibility and applicability of the RAINS-ASIA model. Regional and in-country monitoring programs could generate the data needed for validation. It is essential that such programs be designed to monitor change so that on-the-ground effects of abatement policies and investment measures can be understood. Even as scientific work on validation goes on, the model for policymaking should be applied on the policymaking level through case studies that focus on issues associated with continued growth of energy use and combustion of fossil fuels.

All of these issues require continued collaboration, institution building, and exchange of expertise with Asian institutions. Three principal networks for energy and emissions, atmosphere, and environmental impacts have already been established in Phase I. Asian focal centers for each network have already begun the process of building long-term international networks to develop the model further. Strengthening these existing networks through training, workshops, and distribution of computer equipment and models is a prerequisite to increasing awareness of and interest in the RAINS-ASIA model and its results. In addition, it is important to develop mechanisms for technical and financial assistance to support both institutional development and policy and investment actions.

RAINS-ASIA Phase II

Some of the recommendations made in this chapter have been incorporated in RAINS-ASIA Phase II program, initiated by the World Bank and the Asian Development Bank with donor support, to be implemented during 1997–98. Phase II will focus on three areas:

- Dissemination and training workshops in Asian countries, including distribution of the RAINS-ASIA and RESGEN software, in-country training workshops, and financial support for Asian researchers and scientists to participate in international conferences
- Expansion of the passive sampling program; verification of energy, soil, and vegetation data, including model features for evaluation of direct effects of sulfur dioxide; assessment of model uncertainties and interannual meteorological variability; and some improvements to software
- Creation of a nodal point for the RAINS-ASIA model development in countries that have potential acid rain problems and development of a forum for intercountry dialogue in the Asian region.

APPENDIX
Database Structure of the RAINS-ASIA Model

This appendix contains a map of the subnational regions used in the RAINS-ASIA model, with a key to the regions (table A1), as well as tables showing the economic sectors, fuel types, and control technologies included in the model.

Figure A1. Subnational Regions in the RAINS-ASIA Model

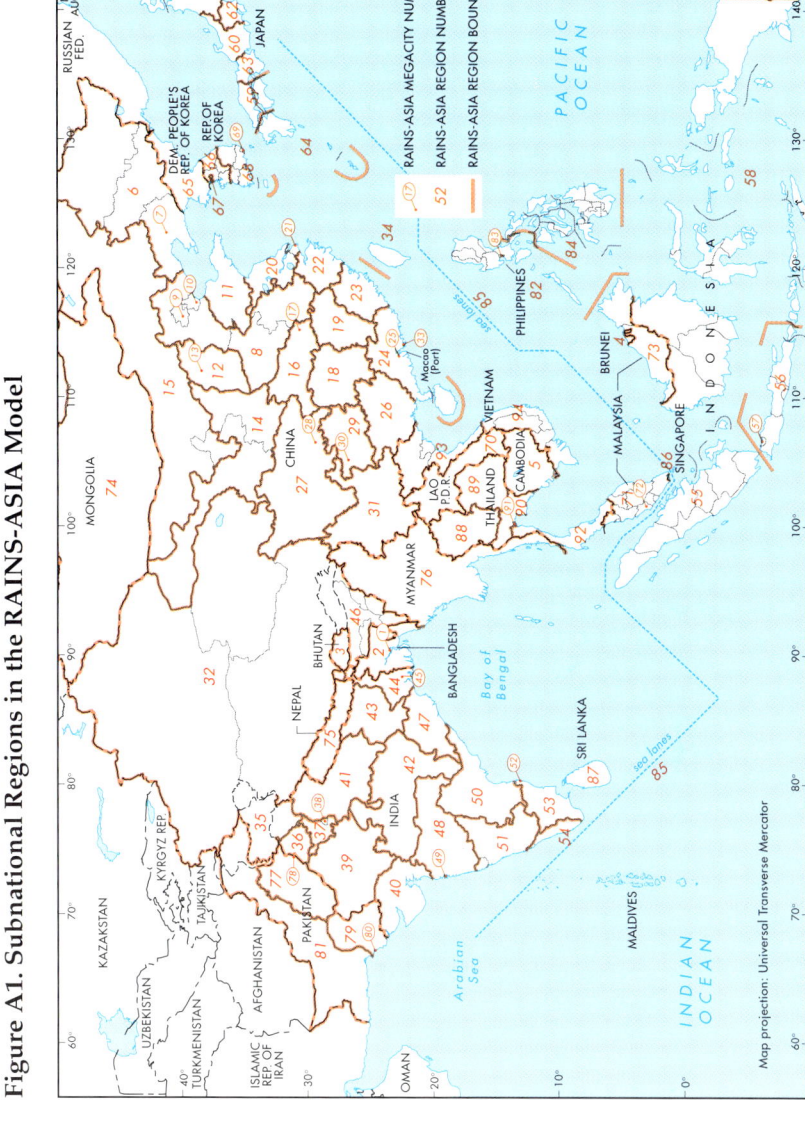

Map projection: Universal Transverse Mercator

Table A1.   Economies and Regions in the RAINS-ASIA Model

Region Number	Country	Country code	Region code	Region comprises
1	Bangladesh	BANG	*DHAK*	*Dhaka*
2			REST	The entire nation except DHAK
3	Bhutan	BHUT	WHOL	The entire nation
4	Brunei	BRUN	WHOL	The entire nation
5	Cambodia	CAMB	WHOL	The entire nation
6	China	CHIN	NEPL	"North-Eastern Plain," comprising Heilongjiang, Jilin, and Liaoning provinces except SHEN
7			*SHEN*	*Shenyang*
8			HEHE	Anhui, Beijing, Hebei, Henen, and Tianjin provinces except BEIJ and TIAN
9			*BEIJ*	*Beijing*
10			*TIAN*	*Tianjin*
11			SHND	Shandong province
12			SNHX	Shanxi province except TAIY
13			*TIAY*	*Taiyuan*
14			SHGA	Gansu and Shaanxi provinces
15			IMON	"Inner Mongolia," comprising Nei Mongol and Ningxia provinces
16			HUBE	Hubei province except WUHA
17			*WUHA*	*Wuhan*
18			HUNA	Hunan province
19			JINX	Jiangxi province
20			JINU	Jiangsu province except SHAN
21			*SHAN*	*Shanghai*
22			ZHEJ	Zhejiang province
23			FUJI	Fujian province
24			GUAH	Guangdong and Hainan provinces except GUAZ and HONG
25			*GUAZ*	*Guangzhou*
26			GUAX	Guangxi province
27			SICH	Sichuan province except CHON
28			*CHON*	*Chongqing*
29			GUIZ	Guizhou province except GUIY
30			*GUIY*	*Guiyang*
31			YUNN	Yunnan province
32			WEST	"West," comprising Qinghai, Xinjiang, and Xizang provinces
33			*HONG*	*Hong Kong, China*
34			TAIW	Taiwan province

Region Number	Country	Country code	Region code	Region comprises
35	India	INDI	WHIM	"Western Himalayas," comprising states of Himachal Pradesh, and Jammu and Kashmir
36			PUNJ	State of Punjab; Chandigarh
37			HARY	State of Haryana except DELH
38			DELH	New Delhi
39			RAJA	State of Rajasthan
40			GUJA	State of Gujarat
41			UTPR	State of Uttar Pradesh
42			MAPR	State of Madhya Pradesh
43			BIHA	State of Bihar
44			BENG	State of West Bengal except CALC
45			CALC	Calcutta
46			EHIM	"Eastern Himalayas," comprising states of Arunachal Pradesh, Assam, Manipur, Meghalaya, Mizoram, Nagaland, Sikkim, and Tripura
47			ORIS	State of Orissa
48			MAHA	State of Maharashtra except BOMB
49			BOMB	(Mumbai) Bombay
50			ANPR	State of Andhra Pradesh
51			KARN	State of Karnataka; Goa
52			MADR	Chennai (Madras)
53			TAMI	State of Tamil Nadu except MADR
54			KERA	State of Kerala
55	Indonesia	INDO	SUMA	Bengkulu, D.I. Aceh, Jambi, Lampung, Riau, Sumatera Barat, Sumatera Seletan, and Sumatera Utara provinces
56			JAVA	Bali, D.I. Yogyakarta, D.K.I. Jakarta, Jawa Barat, Jawa Tengah, and Jawa Timur provinces except JAKA
57			JAKA	Jakarta

(Table continues on the following page.)

Table A1   *(continued)*

Region Number	Country	Country code	Region code	Region comprises
58			REST	Irian Jaya, Klimantan Barat, Kalimantan Selatan, Kalimantan Tengah, Kalimantan Timur, Maluku, Nusa Tenggara Barat, Nusa Tenggara Timur, Sulawesi Selatan, Sulawesi Tengah, Sulawesi Tenggara, Sulawesi Utara, Timor Timur
59	Japan	JAPA	CHSH	Chugoku and Shikoku districts
60			CHUB	Chubu district
61			HOTO	Hokkaido andTohoku districts
62			KANT	Kanto district
63			KINK	Kinki district
64			KYOK	Kyushu and Okinawa districts
65	Korea, Dem. People's Rep. of	KORN	WHOL	The entire nation
66	Korea, Rep. of	KORS	NORT	Kangwon, Kyonggi, North Chungchong, and South Chunchong provinces except SEOI
67			SEOI	Special cities of Inchon and Seoul
68			SOUT	Cheju, North Cholla, North Kyongsang, South Cholla, and South Kyongsang provinces except PUSA
69			*PUSA*	*Pusan*
70	Laos People's Dem. Rep.	LAOS	WHOL	The entire nation
71	Malaysia	MALA	PENM	"Peninsular Malaysia" comprising states of Johor, Kedah, Kelantan, Melaka, Negri Sembilan, Pahang, Perak, Perlis, Pulau Pinang, Trengganu, and Selangor, except KUAL
72			*KUAL*	*Kuala Lumpur*
73			SASA	States of Sabah and Sarawak
74	Mongolia	MONG	WHOL	The entire nation
75	Nepal	NEPA	WHOL	The entire nation
76	Myanmar	MYAN	WHOL	The entire nation
77	Pakistan	PAKI	PUNJ	Punjab province except LAHO

Region Number	Country	Country code	Region code	Region comprises
78			*LAHO*	*Lahore*
79			SIND	Sindh province except KARA
80			*KARA*	*Karachi*
81			NWBA	Balochistan and North West Frontier provinces
82	Philippines	PHIL	LUZO	Cordillera Administrative Region, National Capital Region, Region I, Region II, Region III, and Region IV except MANI
83			*MANI*	*Manila*
84			BVMI	Autonomous Region of Muslim Mindanao, Region V, Region VI, Region VII, Region VIII, Region IX, Region X, Region XI, and Region XII
85	sea lanes	SEAL	SEAL	Shipping routes on the high seas
86	Singapore	SING	WHOL	The entire nation
87	Sri Lanka	SRIL	WHOL	The entire nation
88	Thailand	THAI	NHIG	"North Highlands," comprising the North Region
89			NEPL	"Northeast Plateau," comprising the Northeast Region
90			CVAL	"Central Valley," comprising the Central Region except BANG
91			*BANG*	*Bangkok*
92			SPEN	Southern Peninsula, comprising the South Region
93	Vietnam	VIET	NORT	North Central, Northern Uplands, and Red River Delta planning regions
94			SOUT	Central Coast, Central Highlands, Mekong River Delta, and Southeast planning regions

Note: The Region numbers correspond to the numbers used on the map. Names in italics are RAINS-ASIA megacities. The model also considers international sea (shipping) lanes as a separate region.

Table A2.   Economic Sectors in the RAINS-ASIA Model

	Sector/subsector	Code
1	Fuel production and conversion	CON
	Combustion	CON_COMB
	Losses	CON_LOSS
2	Power plants, district heating	PP
	Existing wet bottom	PP_EX_WB
	Existing other	PP_EX_OTH
	New	PP_NEW
3	Household and other consumers	DOM
4	Transport	TRA
	Road transport	TRA_RD
	Cars, motorcycles, light-duty trucks	TRA_RD_LD
	Two-stroke transport	TRA_RD_LD2
	Four-stroke transport	TRA_RD_LD4
	Heavy-duty vehicles (trucks, buses)	TRA_RD_HD
	Other transport (rail, inland water, and coastal)	TRA_OTHER
5	Industry	IN
	Combustion in boilers for electricity and heat	IN_BO
	Other industrial combustion (furnaces)	IN_OC
	Process emission	IN_PR
	Oil refineries	IN_PR_REF
	Coke plants	IN_PR_COKE
	Sinter plants	IN_PR_SINT
	Pig iron (blast furnaces)	IN_PR_PIGI
	Nonferrous metal smelters	IN_PR_NFME
	Sulfuric acid plants	IN_PR_SUAC
	Nitric acid plants	IN_PR_NIAC
	Cement and lime plants	IN_PR_CELI
	Pulp mills	IN_PR_PULP
6	Non-energy use	NONEN

Table A3.   Fuel Types in the RAINS-ASIA Model

	Fuel type	Code
1	Brown coal or lignite, high grade	BC1
2	Brown coal or lignite, low grade	BC2
3	Hard coal, high quality	HC1
4	Hard coal, medium quality	HC2
5	Hard coal, low quality	HC3
6	Derived coal (coke, briquettes)	DC
7	Other solid-low S (biomass, waste, wood)	OS1
8	Other solid-high S (includes high S waste)	OS2
9	Heavy fuel oil	HF
10	Medium distillates (diesel, light fuel oil)	MD
11	Light fractions: gasoline, kerosene, napthas, LPG	LF
12	Gas (includes other gases)	GAS
13	Renewable (solar, wind, small hydro)	REN
14	Hydro	HYD
15	Nuclear	NUC
16	Electricity	ELE
17	Heat (steam, hot water)	HT

Table A4.   Sulfur Dioxide Emissions Control Technologies in the RAINS-ASIA Model

	Type of control	Code
1	Low-sulfur fuels	LSFUEL
	Low-sulfur coal	LSCO
	Low-sulfur coke	LSCK
	Low-sulfur fuel oil	LSHF
	Low-sulfur medium distillates—stage 1 (0.3 percent sulfur)	LSMD1
	Low-sulfur medium distillates—stage 2 (0.05 percent sulfur)	LSMD2
2	Flue-gas desulfurization	FGD
	Limestone injection	LINJ
	Wet flue-gas desulfurization	WFGD
	Regenerative flue-gas desulfurization	RFGD
3	Process/technology emissions	SO2
	Stage 1 control (50 percent efficiency)	SO2PR1
	Stage 2 control (70 percent efficiency)	SO2PR2
	Stage 3 control (80 percent efficiency)	SO2PR3

Bibliography

Ahmad, J. U. 1991. "Acid Rain in Bangladesh." In Wesley K. Foell and D. Sharma, eds., *Proceedings, Second Annual Workshop on Acid Rain and Emissions in Asia, Nov. 19–22, 1990.* Bangkok: Asian Institute of Technology.

Amann, Markus, and Janusz Cofala. 1995. "Scenarios of Future Acidification in Asia: Exploratory Calculations." In "RAINS-ASIA Technical Report: The Development of an Integrated Model for Sulfur Deposition." World Bank, Asia Technical Group, Washington, D.C.

Amann, Markus, Janusz Cofala, and Leen Hordijk. 1995. "Integrated Assessment." In "RAINS-ASIA Technical Report: The Development of an Integrated Model for Sulfur Deposition." World Bank, Asia Technical Group, Washington, D.C.

Carmichael, G. R., and R. L. Arndt. 1995. "Long-Range Transport and Deposition of Sulfur in Asia." In "RAINS-ASIA Technical Report: The Development of an Integrated Model for Sulfur Deposition." World Bank, Asia Technical Group, Washington, D.C.

Cofala, Janusz, and W. Schopp. 1995. "Assessing Future Acidification in Europe." Note prepared for 15th meeting of the UN ECE Task Force on Integrated Assessment Modelling, The Hague, May 1995. International Institute for Applied Systems Analysis, Laxenburg, Austria.

Dianwu, Zhao, and Zhang Xiaoshan. 1992. "Acid Rain in Southwestern China." In W. K. Foell and C. Green, eds., *Proceedings, Third Annual Workshop on Acid Rain and Emission in Asia, Nov. 18–21, 1991.* Bangkok: Asian Institute of Technology.

Foell, Wesley K., and D. Sharma, eds. 1991. *Proceedings, Second Annual Workshop on Acid Rain and Emissions in Asia, Nov. 19–22, 1990.* Bangkok: Asian Institute of Technology.

Foell, Wesley K., and C. Green, eds. 1992. *Proceedings, Third Annual Workshop on Acid Rain and Emissions in Asia, Nov. 18–21, 1991.* Bangkok: Asian Institute of Technology.

Gian, T. X., N. T. Van, and N. H. Nihn. 1992. "Air Pollution and Acid Rain in Vietnam." In Wesley K. Foell and C. Green, eds., *Proceedings, Third Annual Workshop on Acid Rain and Emissions in Asia, Nov. 18–21, 1991.* Bangkok: Asian Institute of Technology.

Green, C., J. Legler, A. Sarkar, and Wesley Foell. 1995. "Regional Energy Scenario Generation Module." In "RAINS-ASIA Technical Report: The Development of an Integrated Model for Sulfur Deposition." World Bank, Asia Technical Group, Washington, D.C.

Hettelingh, J. P., M. Chadwick, H. Sverdrup, and D. Zhao. 1995. "Impact Module." In "RAINS-ASIA Technical Report: The Development of an Integrated Model for Sulfur Deposition." World Bank, Asia Technical Group, Washington, D.C.

Hong, M. S. 1991. "Description of Acid Rain Problem in Korea." In Wesley K. Foell and D. Sharma, eds., *Proceedings, Second Annual Workshop on Acid Rain and Emissions in Asia, Nov. 19–22, 1990.* Bangkok: Asian Institute of Technology.

Hordijk, Leen, Wesley Foell, and Jitendra J. Shah. 1995. "Introduction." In "RAINS-ASIA Technical Report: The Development of an Integrated Model for Sulfur Deposition." World Bank, Asia Technical Group, Washington, D.C.

National Acid Precipitation Assessment Program. 1991. *1990 Integrated Assessment Report.* Washington, D.C.

Sridharan, P. V., and S. Saksena. 1991. "Description of Acid Rain Problem in India." In Wesley K. Foell and D. Sharma, eds., *Proceedings, Second Annual Workshop on Acid Rain and Emissions in Asia, Nov. 19–22, 1990.* Bangkok: Asian Institute of Technology.

Streets, D., Markus Amann, N. Bhatti, Janusz Cofala, and C. Green. 1995. "Emissions and Control." In "RAINS-ASIA Technical Report: The Development of an Integrated Model for Sulfur Deposition." World Bank, Asia Technical Group, Washington, D.C.

UN ECE. 1994. *Second Sulfur Protocol.* Geneva: United Nations.

For a copy of the RAINS-ASIA model, please contact:

International Institute for Applied Systems Analysis (IIASA)
Attn: Margaret Gottsleben
A2631 Laxenburg, Austria
Phone: +43 2236 807, ext 474
Fax: +43-2236-71313
Email: gottsleb@iiasa.ac.at

Directions in Development

Begun in 1994, this series contains short essays, written for a general audience, often to summarize published or forthcoming books or to highlight current development issues.

(continued on the following page)

Directions in Development (*continued*)